# ATMOSPHERE–OCEAN INTERACTIONS

## VOLUME 2

# International Series on Advances in Fluid Mechanics

## Objectives

The field of fluid mechanics is rich in exceptional researchers worldwide who have advanced the science and brought a greater technical understanding of the subject to their institutions, colleagues and students.

This book series has been established to bring such advances to the attention of the broad international community. Its aims are achieved by contributions to volumes from leading researchers by invitation only. This is backed by an illustrious Editorial Board who represent much of the active research in fluid mechanics worldwide.

Volumes in the series cover areas of current interest and active research and will include contributions by leaders in the field.

Topics for the series include: Bio-Fluid Mechanics, Biophysics and Chemical Physics, Computational Methods for Fluids, Experimental & Theoretical Fluid Mechanics, Fluids with Solids in Suspension, Fluid-Structure Interaction, Geophysics, Groundwater Flow, Heat and Mass Transfer, Hydrodynamics, Hydronautics, Magnetohydrodynamics, Marine Engineering, Material Sciences, Meteorology, Ocean Engineering, Physical Oceanography, Potential Flow of Fluids, River and Lakes Hydrodynamics, Slow Viscous Fluids, Stratified Fluids, High Performance Computing in Fluid Mechanics, Tidal Dynamics, Viscous Fluids, and Wave Propagation and Scattering.

---



---

# Atmosphere–Ocean Interactions

## Volume 2

Edited by

W. Perrie
*Bedford Institute of Oceanography, Canada*

WIT*PRESS* Southampton, Boston

# Ocean–Atmosphere Interactions
## Volume 2

*Series: Advances in Fluid Mechanics, Vol. 39*

Edited by

W. Perrie

*Bedford Institute of Oceanography, Canada*

Published by

**WIT Press**
Ashurst Lodge, Ashurst, Southampton, SO40 7AA, UK
Tel: 44 (0) 238 029 3223; Fax: 44 (0) 238 029 2853
E-Mail: witpress@witpress.com
http://www.witpress.com

For USA, Canada and Mexico

**WIT Press**
25 Bridge Street, Billerica, MA 01821, USA
Tel: 978 667 5841; Fax: 978 667 7582
E-Mail: infousa@witpress.com
http://www.witpress.com

British Library Cataloguing-in-Publication Data

A Catalogue record for this book is available
from the British Library

ISBN: 1-85312-929-1
ISSN: 1353-808X

Library of Congress Catalog Card Number: 2001096403



Printed in Great Britain by Lightning Source UK Ltd., Milton Keynes

# Contents

# Preface

The purpose of this second volume on atmosphere–ocean interactions is to present several of the key mechanisms that are important for the development of marine storms. Understanding atmosphere–ocean interactions has taken on increased importance in recent years, as increased levels of population and human development are reached in coastal areas. Human activities that depend on the oceans, for a variety of interests, require improvements in operational forecasts for marine weather and ocean conditions, and associated marine climate. Factors influencing storm intensity and development are related to small-scale atmosphere–oceanic features. Large-scale climate change may affect environmental factors related to storm climatologies. These topics are presented as separate chapters that are self-contained, to a large extent.

The book consists of eight chapters. Chapters 1–5 are concerned with marine observations and understanding their parameterizations as they relate to atmosphere–ocean systems. Chapters 6–8 consider some of the implications of these parameterizations, as related to applications in coupled atmosphere, ocean, and wave model systems.

Chapter 1 presents a review of the state of the art of parameterizations for air–sea turbulent fluxes. A data set of eddy-correlation fluxes compiled during eight field experiments, and representing a wide range of conditions, is used to investigate traditional bulk flux parameterizations (i.e., wind-speed dependent momentum coefficient, constant sensible and latent heat coefficients), including the effects of sea state. Chapter 2 examines the influence of waves on surface stress and related parameterizations of momentum roughness length. Wave-related impacts on the near surface atmospheric boundary conditions are also considered. In light of laminar–turbulent regimes at the air–sea interface related to local wind and wind waves, Chapter 3 considers the challenge of deriving a better parameterization for the CO2 transfer between the atmosphere and the oceans. Consideration is given to nonlinear coupling of waves with local wind drift and turbulence, caused by the wind and wind-wave breaking at the air–sea interface. Chapter 4 reviews the marine meteorological capacities of synthetic aperture radar (SAR), particularly the attributes of its potential in the study of air–sea interaction. Examples include marine meteorological phenomena routinely imaged by SAR such as microscale cellular convection, microscale roll vortices, microscale gravity waves, mesoscale gravity waves, mesoscale convection, polar mesoscale cyclones, tropical cyclones, macroscale fronts, and extratropical cyclones. Chapter 5 reviews the three-dimensional current and temperature structures excited by the passage of tropical cyclones, over the scales of the atmospheric forcing. Relations between the internal wave fluxes and atmospheric winds are shown to be consistent with established observations and parameterizations of the air–sea momentum flux. Chapter 6 applies established parameterizations for air–sea fluxes of heat and momentum to coupled ocean–atmosphere wave simulations for hurricanes and investigates the sensitivity of estimates of air–sea $CO_2$ exchange rates as influenced by storm-related processes. Chapter 7 considers extratropically transitioning tropical cyclones beginning from their tropical formation, the conditions under which they evolve, and their spatial and temporal distributions. The discussion includes the need for synoptic support for the initial intensification of tropical cyclones, followed by a strong energy source as transition occurs. Chapter 8 estimates changes of wave heights, based on observed and possible future changes in the North Atlantic and in the North Pacific. Both oceans had significant changes in both winter and fall seasonal means and extremes of significant wave heights during 1958–1997; they are also projected to have significant changes in the 21st century. The rate and sign of the projected future changes appear to be quite dependent on the forcing conditions, and the rate of increase in the greenhouse gases forcing.

I am very thankful to my colleagues for their invaluable contributed chapters, without which this book would not be possible, to Drs. Peter Smith and John Loder at BIO for encouraging me to bring together this volume, to Prof. Matiur Rahman at Dalhousie University for supporting me in the preparation of this volume, and to the Panel on Energy Research and Development of Canada for providing funding support for this project. I would like to dedicate the volume to my wife Chi and family for their continuing patience, and to the memory of my parents, John and Clara Perrie.

W. Perrie
Dartmouth, Nova Scotia
2006

# CHAPTER 1

# On parameterizations of air–sea fluxes

W.M. Drennan
*Division of Applied Marine Physics, Rosenstiel School of Marine and Atmospheric Science, University of Miami, USA.*

## Abstract

In order to understand and model many physical processes relating to the atmosphere and ocean, it is necessary to consider how the two are coupled. The air–sea fluxes of momentum, heat, and mass are the key quantities linking the two fluids. However, since fluxes over the sea are difficult and costly to measure, especially over the time and space scales required for model input, they have long been parameterized in terms of more readily available mean parameters and bulk exchange coefficients.

In this paper we review the present state of the art of parameterizing air–sea turbulent fluxes. A data set of eddy-correlation fluxes compiled during eight field experiments, and representing a wide range of conditions, is used to investigate traditional bulk flux parameterizations (i.e. wind-speed dependent momentum coefficients, constant sensible and latent heat coefficients), as well as several more recent algorithms. Particular emphasis is given to the effect of sea state on the fluxes.

## 1 Introduction

Over the past decades, our ability to model the global environment has improved considerably. This is not only due to vastly superior computer resources but also due to a better understanding of many of the physical processes. The focus of recent years has been on the longer term, i.e. climate models (Shindell *et al.* [1]), and also extreme events, such as hurricanes (Bao *et al.* [2]). In both cases, progress relies on treating the atmosphere and ocean as a coupled system, taking into account exchanges at the air–sea interface. Here we focus on the turbulent exchanges, referring the reader elsewhere (e.g. Garratt [3]) for information on radiative transfers.

Turbulence at the air–sea interface is generated primarily by two mechanisms: buoyancy and shear, the latter generated by the roughness at the air–sea interface. The sea surface is characterized by waves with wavelengths ranging over five orders of magnitude, from centimeter scale ripples to ocean swells hundreds of meters long. The composition of the wave field varies with time and space primarily according to the surface wind field and its history, and also surface currents, air and sea temperatures, and, near the coast, bathymetry. The air–sea interface is fundamentally different from the air–land interface in that the roughness elements are mobile, and evolve with both the atmosphere and the ocean. The waves grow by extracting momentum from the air through pressure–slope correlations. As a wave field develops, the surface roughness changes, which in turn affects the turbulent momentum transfer. The waves and atmospheric boundary layer represent a truly coupled system.

Since air–sea fluxes are the boundary conditions for both the atmosphere and the ocean, knowledge of them at suitable spatial and temporal scales is important. However, direct measurement of these fluxes over the sea is difficult and expensive, and hence measured fluxes are rarely available. Instead, models rely on parameterized fluxes, based on more easily measured mean variables such as wind speed, air and sea temperatures, etc. The development of accurate flux parameterizations has been a goal of air–sea interaction research for many decades. Flux parameterizations have progressed well beyond the early efforts where the bulk transfer coefficients were assumed to be constants.

Here we assess the present state of air–sea turbulent flux parameterizations, with particular attention paid to the effects of sea state. We start by summarizing some of the theory underlying boundary layer turbulence (Section 2). We then introduce an extensive data set combining turbulent flux and wave measurements from a wide variety of conditions (Sections 3 and 4), and use it to study the effect of sea state on the momentum (Section 5) and heat fluxes (Section 6). Finally, we discuss limitations to our current knowledge, with particular emphasis on the high wind regime (Section 7).

## 2 Turbulent fluxes and the Monin–Obukhov similarity theory

The turbulent fluxes of momentum, sensible heat, and mass are given by

$$\vec{\tau} = \rho(-\overline{w'u'}\hat{i} - \overline{w'v'}\hat{j}), \tag{1a}$$

$$H = \rho C_P \overline{w'\theta'}, \tag{1b}$$

and

$$M_X = \rho\overline{w'x'}, \tag{1c}$$

respectively, where the primes indicate turbulent fluctuations, $u$, $v$, $w$, and $\theta$ represent up-wind velocity, cross-wind velocity, vertical velocity, and potential temperature, respectively, $x$ is the mixing ratio of some constituent with respect to dry air,

$\rho$ is the air density, $C_P$ is the specific heat at constant pressure, and the overbars represent averages over a time period of $O(20\,\text{min})$.

The mass fluxes of particular current interest are those of humidity $q$, and carbon dioxide. Although eqn (1c) is written in terms of the mixing ratio, typical sensors measure the mass concentration, $c = \rho x$. Webb *et al.* [4] show that if the mass concentration is used in eqn (1c), a correction must be made to account for the fluctuations in $\rho$ caused by temperature and water vapor. The second order correction is given by

$$M_X = \overline{w'c'} - \bar{c}T_{\text{v}}^{-1}\overline{w't'_{\text{v}}}, \tag{2}$$

where the virtual temperature $t_{\text{v}} = \theta(1+0.61r)$, where $r$ is the mixing ratio of water vapor in dry air. The upper case variables ($T_{\text{v}}$, $U$, etc.) refer to mean quantities. For $CO_2$ flux, the Webb correction is very important, being of the same order as the measured flux (Fairall *et al.* [5]). For water vapor, the Webb correction is small, typically several percent, and is usually neglected. The latent heat flux $E$ is related to the humidity flux $M_q$ by $E = L_{\text{v}}M_q$, where $L_{\text{v}}$ is the latent heat of vaporization.

Efforts to model the turbulent exchange rely on the ideas of flow similarity, as proposed by Monin and Obukhov [6]. Under the assumptions of stationarity and homogeneity, it can be shown (Busch [7]) that the heat flux is constant with height and that the vertical gradient of the momentum flux is balanced by Coriolis terms. Close to the surface, typically taken as the lower 10% of the boundary layer, the effect of the Coriolis terms is small, and the momentum flux is usually assumed to be constant with height. Although the error associated with this constant-flux assumption is small near the surface, it increases linearly with height. It is clearly significant for most aircraft measurements, and may also be at the 20 m measurement heights typical of ships. Hence the correction of Donelan [8] should be applied.

Within the "constant" flux layer, one can define the turbulence scaling parameters,

$$u_* = (|\tau|/\rho)^{1/2}, \quad t_* = -(H/\rho C_P)/u_* \quad \text{and} \quad x_* = -(M_X/\rho)/u_*, \tag{3}$$

for momentum, heat, and mass, respectively. $u_*$, defined in eqn (3), is known as the friction velocity. Then, assuming that turbulence in the surface layer is generated by a combination of surface shear and buoyancy (this neglects the region very near the surface, where molecular processes are also important), the Monin–Obukhov (MO) theory asserts that the mean gradients and scaling parameters are related through universal dimensionless gradient functions $\Phi$ of the form:

$$\frac{\partial U}{\partial z} = \frac{u_*}{\kappa z}\Phi_u\left(\frac{z}{L}\right); \frac{\partial \Theta}{\partial z} = \frac{t_*}{\kappa z}\Phi_\theta\left(\frac{z}{L}\right); \frac{\partial X}{\partial z} = \frac{x_*}{\kappa z}\Phi_x\left(\frac{z}{L}\right), \tag{4}$$

where $z$ is the mean height above the surface, and $L$ is the Obukhov length,

$$L = -u_*^3[\kappa g(H/(C_P\Theta) + 0.61E/L_{\text{v}})/\rho]^{-1} = -u_*^3[\kappa F_{\text{B}}]^{-1}, \tag{5}$$

where $F_{\text{B}}$ is the buoyancy flux, $\kappa \approx 0.4$ is the von Kármán constant, and $g$ is the gravitational constant. The Obukhov length represents the height where turbulence production due to shear is equal to that due to buoyancy.

Integrating eqn (4) from the surface to some height $z$ in the constant flux layer gives

$$\frac{U(z)-U(0)}{u_*}=\frac{\Phi_u(0)}{\kappa}\left[\log\left(\frac{z}{z_0}\right)-\Psi_u\left(\frac{z}{L}\right)\right], \tag{6a}$$

$$\frac{\Theta(z)-\Theta(0)}{t_*}=\frac{\Phi_\theta(0)}{\kappa}\left[\log\left(\frac{z}{z_{0t}}\right)-\Psi_\theta\left(\frac{z}{L}\right)\right], \tag{6b}$$

$$\frac{X(z)-X(0)}{x_*}=\frac{\Phi_x(0)}{\kappa}\left[\log\left(\frac{z}{z_{0x}}\right)-\Psi_x\left(\frac{z}{L}\right)\right], \tag{6c}$$

where $z_0$, $z_{0t}$, and $z_{0x}$ are the surface roughness lengths for momentum, heat, and mass (i.e. the heights where the profiles cross zero). The functions $\Psi$ in eqn (6) are related to the integrated flux profile or gradient relations $\Phi$. Recent data support $\Phi_u(0)=1$, while values of 0.85 (Donelan [8]) and 0.95 (Högström [9]) have each been recommended for $\Phi_\theta(0)=\Phi_x(0)$. Note that it is common to replace $\Phi_\theta(0)$ and $\Phi_x(0)$ in eqn (6) by the turbulent Prandtl number, $\mathrm{Pr}=\Phi_\theta(0)/\Phi_u(0)=\Phi_x(0)/\Phi_u(0)$.

In neutral conditions, where buoyant forcing becomes negligible, $|z/L|\to 0$, and the $\Psi$ functions approach zero. Equation (6) then yields the well-known logarithmic forms,

$$\frac{U_\mathrm{N}(z)-U(0)}{u_*}=\frac{1}{\kappa}\log\left(\frac{z}{z_0}\right), \tag{7a}$$

$$\frac{\Theta_\mathrm{N}(z)-\Theta(0)}{t_*}=\frac{\mathrm{Pr}}{\kappa}\log\left(\frac{z}{z_{0t}}\right), \tag{7b}$$

$$\frac{X_\mathrm{N}(z)-X(0)}{x_*}=\frac{\mathrm{Pr}}{\kappa}\log\left(\frac{z}{z_{0x}}\right), \tag{7c}$$

where the subscript N represents neutral conditions.

Since the introduction of MO theory, considerable work has been carried out to determine the dimensionless gradient functions. Early experimental verification of MO similarity in the atmosphere over land was given by Zilitinkevich and Chalikov [10], Dyer and Hicks [11], and Businger *et al.* [12]. Although there is some scatter between the various data sets and the resultant profile functions, Högström [9, 13] found the various forms to be consistent within 10–20% within the region $|z/L|<0.5$.

The first systematic investigations of the equivalent problem in the marine environment were made around the same time. In general, flux measurements over the ocean showed good spectral agreement with the terrestrial measurements (e.g. Miyake *et al.* [14]; Pond *et al.* [15]). An exception was found in the presence of swell where upward momentum fluxes were occasionally noted (Volkov [16]; Makova [17]). Likewise, Holland *et al.* [18] reported negative wind speed gradients (larger winds near the surface) in swell conditions.

Pond *et al.* [15], Paulson *et al.* [19], and Davidson [20] presented simultaneous measurements of fluxes, profiles, and surface waves from the FLIP platform during the BOMEX experiment. These data show general support for the MO theory. However, as the motion of the platform was not measured, it was not possible to separate wave-induced motion of the platform from a possible wave-coherent signal in the velocity components: all anomalous signals in the motion band were filtered out. Although necessary at the time, a consequence of this is the loss of wave-coherent information (crucial for the study of wave effects). To summarize, most early experiments, except for some carried out in swell-dominated conditions, were seen to support MO theory over water, with most data indicating general support for the land-based stability functions.

## 3 Measuring turbulent fluxes

### 3.1 Direct methods

The calculation of fluxes via eqn (1) requires the measurement of the turbulent fluctuations of vertical velocity $w'$, along with $u', v', \theta'$, and $x'$. This direct calculation of fluxes is referred to as the "eddy correlation" or "direct covariance" method. Sampling of the turbulent signals must be fast enough, and for a sufficiently long period, to capture all scales contributing to the flux. Based on the universal cospectral shapes of Kaimal *et al.* [21], in neutral conditions 95% of the cospectral energy of momentum (heat) is contained in the dimensionless frequency range 0.0018 (0.0024) $< fz/U <$ 1.3 (2.5), where $f$ is the frequency [Hz]. Consequently, for typical measurement heights and wind speeds, sampling frequency and time series length of orders 10 Hz and 20 min, respectively, are recommended. The sampling length should be increased if strongly unstable conditions ($z/L \ll 0$) are expected, and may be reduced in very stable conditions. Sampling lengths in excess of $O$(60 min) are not recommended due to the influence of mesoscale effects which do not contribute to the turbulent fluxes. See Mahrt *et al.* [22] for a discussion of this and other flux sampling issues.

Although common over land, direct flux measurements over the sea have been limited by both the mobility and the bulkiness of typical marine platforms. Flux towers are relatively rare in the marine environment, being limited by design constraints to the coastal regions. Instead, fluxes are usually measured from ships, buoys, or aircraft, all mobile platforms. The technology for measuring fluxes from aircraft was developed during the 1960s. Such measurements rely on nose-mounted gust or pressure probes, with corrections for aircraft motion as measured by an inertial platform (e.g. Axford [23]). With the development of compact and inexpensive rate gyros and accelerometers, similar systems have been developed for use on buoys (Anctil *et al.* [24]) and ships (Edson *et al.* [25]). Consequently, the problems associated with platform mobility have essentially been solved.

The second limitation is more serious. Flow distortion, a concern with all turbulence measurements (Wieringa [26]; Wyngaard [27]), can be particularly bad over the sea, where platforms must be designed to survive the hostile marine environment. In considering flow distortion the mean and turbulent components of the flow field are usually considered separately. The effects of mean flow distortion can be assessed through numerical (turbulence closure) models using commercial software (e.g. Yelland *et al.* [28]; Dupuis *et al.* [29]) or flume studies using scale models. Mean flow distortion is characterized by the lifting, tilting, and compression of flow streamlines as they pass over or around the platform. The results of Dupuis *et al.* [29] for flow around the 85 m vessel *L'Atalante* are typical: for moderate, bow-on winds, the flow at the top of a 17 m mast near the bow was decelerated by 8%, tilted upward by ~7°, and uplifted by 1.2 m. The effects vary with wind speed and direction, with the distortion increasing significantly for flow angles greater than 30° relative to the bow.

The effect of distortion of the turbulent components remains beyond the reach of current numerical models, and is therefore more difficult to assess. Recent studies comparing turbulent fluxes measured on large ships with those from nearby buoys (which are assumed to be largely free of flow distortion effects) indicate a good comparison between scalar fluxes, but an enhancement of the ship-board momentum flux by an order of 15% (Edson *et al.* [25]; Pedreros *et al.* [30]). It is expected that agreement is better in conditions with low ship motion, and will worsen if considerable slamming is experienced.

### 3.2 Indirect methods—inertial dissipation

The "inertial dissipation" (ID) method was used by Pond *et al.* [15] as a means to avoid the problems of ship motion and turbulent flow distortion. For momentum flux, the method is derived from the turbulent kinetic energy (TKE) conservation equation which for stationary and horizontally homogeneous conditions is:

$$u_*^2 \frac{\partial U}{\partial z} + F_{\mathrm{B}} - \frac{\partial}{\partial z}\overline{w'e'} - \frac{1}{\rho}\frac{\partial}{\partial z}\overline{w'p'} - \varepsilon = 0, \tag{8}$$

where $e' = (u'^2 + v'^2 + w'^2)/2$ and $p'$ represent fluctuations in TKE and pressure, respectively, and $\varepsilon$ is the TKE dissipation rate. The terms in eqn (8) are, respectively, production of TKE from shear, production (or loss) due to buoyancy, transport of TKE, pressure transport, and loss due to dissipation. Similar expressions based on scalar variance budgets exist for the scalar fluxes (e.g. Large and Pond [31]).

Dividing eqn (8) by $u_*^3/\kappa z$ we arrive at

$$\Phi_u - \zeta - \Phi_e - \Phi_p - \frac{\kappa z}{u_*^3}\varepsilon = 0, \tag{9}$$

where

$$\Phi_e = \frac{\kappa z}{u_*^3}\frac{\partial \overline{w'e'}}{\partial z} \quad \text{and} \quad \Phi_p = \frac{\kappa z}{\rho u_*^3}\frac{\partial \overline{w'p'}}{\partial z}.$$

Using the assumptions of MO theory, the terms in the above equation are expected to be universal functions of $\zeta = z/L$. Then from a known dissipation rate and known dimensionless functions (e.g. those of Högström [13]), the stress can be readily determined. In practice, $\varepsilon$ is usually estimated from the inertial subrange (ISR) of the frequency spectrum, assuming Taylor's hypothesis:

$$\varepsilon = \left(\overline{f^{5/3}S(f)}/\alpha_K\right)^{3/2} (2\pi/U_a), \tag{10}$$

where $S(f)$ is the velocity spectrum, and the mean is taken over frequencies in the ISR, where $S(f) \propto f^{-5/3}$. Here $\alpha_K \approx 0.52$ is the Kolmogorov constant and $U_a$ is the advection velocity of turbulence past the probe. This dependence only on ISR frequencies ($f >\sim 2\,\text{Hz}$) is the key advantage of the ID method. These high frequencies are unaffected by either turbulent (but not mean) flow distortion or platform motion. Hence the method remains popular for flux measurements from vessels (e.g. Yelland *et al.* [28]).

With little experimental data available on either $\bar{\Phi}_e$ or especially $\bar{\Phi}_p$, they are usually combined as a single transport (or imbalance) term $\bar{\Phi}_t = \bar{\Phi}_e + \bar{\Phi}_p$. This term is then either assumed to be zero (Large and Pond [32]) or determined empirically (Dupuis *et al.* [33]). See Sjöblom and Smedman [34] for a critical assessment of these assumptions in marine boundary layers.

Equation (9) can be rewritten to yield the expression for $u_*$:

$$u_* = [\kappa z \varepsilon / (\Phi_u(\zeta) - \zeta - \Phi_t(\zeta))]^{1/3}. \tag{11}$$

The ID heat flux is given by $\overline{w'\theta'} = [\kappa z\, u_*\, \varepsilon_\theta / \Phi_\theta(\zeta)]^{1/2}$, where $\varepsilon_\theta = \left(\overline{f^{5/3}S_\theta(f)}/\alpha_\theta\right) \varepsilon^{1/3} (2\pi/U_a)^{2/3}$. Here the Obukhov–Corrsin constant $\alpha_\theta \approx 0.8$. Expressions are similar for other scalar fluxes (Dupuis *et al.* [29]).

Given the assumptions underlying the ID method, its applicability has been questioned in certain conditions, namely in developing seas (Janssen [35]; but see also Taylor and Yelland [36]), in swell-dominated conditions (Donelan *et al.* [37]), and in light winds (Yelland *et al.* [28]). However, in the mixed seas typical of the open ocean ID fluxes compare well with direct measurements (Dupuis *et al.* [29]).

### 3.3 Indirect methods—profile

The profile method uses the dimensionless gradient functions (4), along with measured profiles, to estimate $u_*, t_*, q_*$, etc., and therefore, via eqn (3), the fluxes. The method is rarely employed over the sea, as profile measurements from vessels are particularly prone to mean flow distortion errors, since the flow distortion around a vessel is itself a function of height. Nevertheless, the recent profile measurements of $CO_2$ flux by McGillis *et al.* [38] are consistent with simultaneous direct flux measurements.

### 3.4 Indirect methods—bulk

Bulk fluxes are not measured fluxes, but estimates of fluxes derived from measured mean variables along with empirical bulk coefficients. Bulk transfer coefficients relate the turbulent fluxes through the scaling parameters (3) to the more easily measured mean quantities $U, \Theta$, and $X$, at the surface (subscript 0), and at some height above the surface. For momentum (the drag coefficient $C_D$), sensible heat (Stanton number $C_H$), and mass (constituent $x$, $C_X$), the $z$-m bulk relations are:

$$C_D(z) = |\tau/\rho| / (U_z - U_0)^2 = (u_*/(U_z - U_0))^2, \tag{12a}$$

$$C_H(z) = -\overline{w'\theta'} / \left[(\Theta_z - \Theta_0)(U_z - U_0)\right], \tag{12b}$$

$$C_X(z) = -M_x / \left[\rho(X_Z - X_0)(U_Z - U_0)\right]. \tag{12c}$$

For humidity, the bulk mass coefficient is known as the Dalton number, $C_E$.

Given the difficulty of measuring fluxes at sea many flux applications, particularly those requiring dense spatial information, such as atmospheric or climate modeling, rely on bulk fluxes. Consequently, considerable efforts have been expended over the past decades in determining accurate bulk parameterizations. A summary of recent efforts is found in Brunke *et al.* [39].

It follows from eqns (7) and (12) that the neutral $z$-m bulk coefficients are directly related to the surface roughness lengths, i.e. $C_{DN}(z) = f(z/z_0)$. Based on dimensional reasoning, Charnock [40] proposed that $z_0 = \alpha u_*^2/g$, where $\alpha$ is known as the Charnock parameter. Charnock took $\alpha$ as a constant, which implies $C_{DN}(z) = f(z, U_N(z))$. Mean experimental values of $\alpha$ lie in the range 0.011–0.018 (Kraus and Businger [41]).

Henceforth, we consider the bulk coefficients evaluated at the standard reference height $z$ of 10 m, i.e. $C_{D10N} = f(U_{10N})$, although for convenience the subscript 10 is usually omitted. This form of the bulk coefficients has been studied for decades, and many different relationships have been determined (see Geernaert [42]). However, it is now evident that no single parameterization of this type is able to adequately model all data sets, nor to describe the variability (beyond sampling error, cf. Sreenivasan *et al.* [43]) within most data sets. One goal here is to determine whether, and under what conditions, such bulk relationships are valid. We first introduce the data sets we will use.

## 4 Field data

The data sets, listed in Table 1, include a wide range of conditions from open ocean to near-shore. Each data set includes eddy correlation measurements of momentum flux along with mean meteorological and sea state parameters.

Three of the experiments are tower based: the Water Air Vertical Exchange Study (WAVES), 1 km off shore in Lake Ontario (Donelan *et al.* [44]; Drennan *et al.* [45]); the Risø Air–Sea Exchange experiment (RASEX), in the Baltic Sea, 2 km offshore (Johnson *et al.* [46]); Humidity Exchange Over the Sea experiment (HEXOS), 9 km offshore in the North Sea (Smith *et al.* [47]; DeCosmo *et al.* [48]). Although HEXOS was focused on humidity flux measurements, these data are not readily available. We use here only the momentum flux data as tabulated by Janssen [49].

Table 1: Summary of field data used in this paper.

| Experiment | Reference | Location | Dates | Platform | Depth (m) | No. of runs (WSD) | Avg. (min) | U (m/s) | Fluxes |
|---|---|---|---|---|---|---|---|---|---|
| AGILE | [50] | Lake Ontario | 1995–1996 | Ship | 4–50 | 101 (82) | 20–30 | 3–15 | U, T, Q, C |
| AWE | | Coastal Atlantic | 2000 | Buoy | 20 | 565 (335) | 60 | 2–13 | U, $T_s$ |
| FETCH | [53] | Mediterranean Sea | 1998 | Buoy | 100 | 853 (323) | 28 | 2–19 | U, $T_s$ |
| GASEX | [55] | Pacific Ocean | 2001 | Buoy | 4000 | 456 (0) | 30 | 3–11 | U, T, Q, C |
| HEXOS | [49] | North Sea | 1986 | Tower | 18 | 50 (50) | 20 | 9–20 | U, Q |
| RASEX | [46] | Baltic Sea | 1994 | Tower | 4 | 80 (80) | 30 | 4–16 | U |
| SWADE | [37] | Coastal Atlantic | 1991 | Ship | 20–1000 | 122 (43) | 17 | 4–14 | U, T, Q |
| WAVES | [45] | Lake Ontario | 1987 | Tower | 12 | 238 (156) | 30 | 2–16 | U |

The quantities in brackets in the "No. of runs" column refer to the number of wind sea dominant (WSD) runs. The column "Avg." gives the average time for calculating the fluxes. The symbols in the "Fluxes" column refer to the availability of fluxes of momentum (U), sensible heat (T), humidity (Q), virtual or sonic temperature ($T_s$) and $CO_2$ (C).

Two other data sets are from small ships: the 15 m R/V *Agile* in Lake Ontario (Donelan and Drennan [50]) and the 20 m SWATH ship R/V *Frederick G. Creed* during the Surface Wave Dynamics Experiment (SWADE), in the coastal Atlantic (Katsaros *et al.* [51]; Donelan *et al.* [37]). Both ships were equipped with fast response dry- and wet-bulb thermocouples, as well as anemometers, yielding momentum, sensible heat, and moisture fluxes. The *Agile* also carried a LICOR 6262 closed path infrared gas analyzer, allowing for the calculation of $CO_2$ flux.

The remaining data sets were collected during several recent deployments of Air–Sea Interaction Spar (ASIS) buoys (Graber *et al.* [52]). Of these deployments, only FETCH (a French acronym for fluxes, sea state and remote sensing in variable fetch conditions) in the Mediterranean Sea has been described elsewhere (Drennan *et al.* [53]); the other two will be summarized briefly here.

The Adverse Weather Experiment (AWE), took place off Florida's Atlantic coast, during April–May 2000. The goal of AWE was to investigate the impact of adverse weather fronts on mixing and air–sea interaction in the shallow water column. The principal measurements during the experiment were TKE dissipation rates from an autonomous underwater vehicle (Dhanak *et al.* [54]), surface currents from an Ocean Surface Current Radar, and air–sea fluxes of momentum and buoyancy, along with surface waves, from an ASIS buoy. The ASIS flux measurements were taken using a Gill R2A sonic anemometer mounted at 7 m above mean sea level. The ASIS buoy was moored at 20 m depth, 2 km from the coast, usually just inshore of the Gulf Stream.

During February 2001, an ASIS buoy was deployed as part of the GASEX experiment which focused on air–sea $CO_2$ exchange in the equatorial Pacific (McGillis *et al.* [55]). In this Lagrangian study, the ASIS buoy was drogued and drifted in the South Equatorial Current. The air–sea fluxes of momentum, heat, water vapor, and $CO_2$ were measured from ASIS buoys (and from the R/V *Ronald H. Brown*) using Gill R2A sonic anemometers and LICOR-7500 open-path gas analyzers (for $CO_2$ and $H_2O$). Here we use the 12 day ASIS data set which includes surface waves and other supporting parameters.

The procedure for calculating fluxes is:

- $w$, $u$, $v$, $\theta$, and $x$ along with surface elevation $\eta$ and, if necessary, platform motion are sampled at 10–20 Hz;
- pressure, relative humidity, sea surface temperature, and surface currents are sampled at $\sim$ 1 Hz;
- signals are quality controlled, broken into $\sim$ 20 min blocks and detrended;
- if necessary, corrections are made to account for instrument response and offset in position (see Katsaros *et al.* [51]);
- on nonstationary platforms, $w$, $u$, $v$, and $\eta$ are corrected to account for platform motion (Anctil *et al.* [24], Drennan *et al.* [56]);
- wind vector is rotated into mean wind direction ($\bar{v} = 0$); tilt correction ($\bar{w} = 0$);
- fluxes are calculated from eqn (1); eqn (2) for $H_2O$ and $CO_2$ fluxes;
- Obukhov length is calculated from eqn (5) using measured $u_*$, and measured (if available) or bulk heat fluxes;

- neutral 10 m mean values $(U, \Theta, X)$ are calculated using Donelan [8] profile relations;
- bulk coefficients are calculated; $U_0$ is neglected except in strong current environments; $q_0$ is calculated assuming saturated air at $\Theta_0$, with a 2% correction for salinity effects when applicable.

Sea state parameters are also calculated. The one-dimensional parameters include: significant wave height $H_s$ (four times the standard deviation of $\eta$), peak frequency $f_p$, peak phase speed $c_p$, and peak wavelength $L_p$, with the latter three being related through the dispersion relationship. In addition, directional wave spectra are calculated using data from multiple wave sensors. Here we use arrays of surface elevation gauges; wave buoys provide similar data—see Pettersson *et al.* [57]. Finally, a wave-partitioning algorithm (e.g. Gerling [58]) is used to separate the wave field into its components. For each wave train, the energy, peak frequency, and peak propagation direction are calculated.

The sea state is classified based on the wind and wave parameters. At most one wave component is designated wind sea, the most energetic component meeting the following criteria: (i) $U_{10N} \cos(\theta_d)/c_p > 0.83$, where $\theta_d$ is the angle between the mean wind and waves (Donelan *et al.* [59]), and (ii) $|\theta_d| < 45°$. Any other wave components are classified as swell. A sea is termed wind sea dominant (WSD) if the energy of the wind sea $E$(ws) exceeds that of the swell(s), $E$(swell). If $E(\text{ws}) > 5E(\text{swell})$, the sea is denoted pure wind sea (PWS). Similar definitions are used for swell-dominated and pure swell conditions.

Figure 1 shows a plot of a compilation of eddy correlation momentum flux data (2464 points) from the eight field experiments. The Smith [60] bulk relation,

$$C_{DN} = (0.61 + 0.063U_{10N}) \times 10^{-3}, \tag{13}$$

shown in Fig. 1, is typical of many drag coefficient curves. The Smith curve is seen to agree reasonably well with the mean of the data at most wind speeds. The drag coefficient as calculated assuming a constant Charnock parameter of 0.015, the mean for these data in the range $5 \le U \le 16$ m/s, provides a slightly improved agreement. However, it is clear that: (i) at moderate to high winds there are a considerable number of points which are much larger than predicted; (ii) at low winds the data are systematically larger than predicted. We discuss each of these points below.

## 5 Momentum flux

### 5.1 Light winds: gustiness and smooth flow

From Fig. 1 it is evident that the highest degree of scatter occurs at low winds. There are several reasons for this. In the free convection limit, as $U \to 0$, shear becomes unimportant, and the friction velocity is no longer a useful scaling variable. It is, however, frequently observed that low wind speeds are often associated

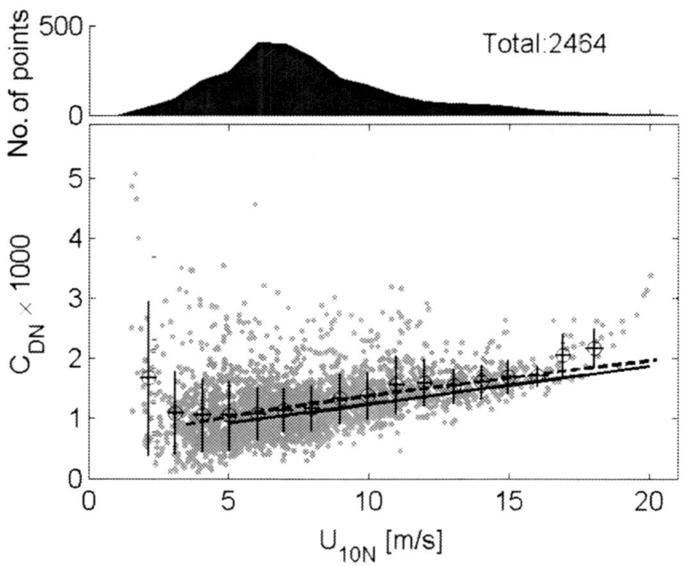


Figure 1: 10 m neutral drag coefficient versus 10 m neutral wind speed for eight field experiments. Also shown are the bulk relation of Smith [60] (solid line) and the drag relation calculated with a constant Charnock parameter of 0.015 (dashed line). The circles represent data averaged in wind speed bins of 1 m/s, showing 1 standard deviation in $U_{10N}$ and $C_{DN}$. The wind speed distribution of the data is shown in the top panel.

with high variability in both speed and direction. This gustiness is associated with large-scale coherent structures in the convective boundary layer (CBL). Businger [61] introduced a new velocity scale, the convective velocity scale, $w_* = (F_B z_i)^{1/3}$ to account for the intermittent effects of gustiness. Here $z_i$ is the CBL height. In light wind conditions, Godfrey and Beljaars [62] proposed replacing the mean wind speed with an effective wind speed ($U_E$), taking into account the gustiness: $U_E^2 = U^2 + u_g^2$, where $u_g = \beta w_*$, and $\beta \approx 1.25$ is a constant. According to Grachev and Fairall [63], gustiness is expected to lead to significant departures from the MO theory for $u_g/U > 0.5$, and must be taken into account in these conditions. During the TOGA-COARE experiment, when winds were below 4 m/s roughly half the time, the convective threshold was met 7.5% of the time (Grachev and Fairall [63]).

The Charnock roughness parameterization assumes the surface roughness to be due entirely to the presence of small-scale gravity waves. This is known as the "rough flow" regime (roughness Reynolds number, $Re_* = u_* z_0/\nu > 2.2$), and is valid for most wind speeds. However at low winds, the viscosity $\nu$ also plays a role in supporting the surface stress. In smooth flow conditions ($Re_* < 0.13$), the roughness length is given by $z_0 = 0.11 u_*/\nu$ (Kraus and Businger [41]). For intermediate Reynolds numbers, Smith [64] proposed the roughness to be the sum of Charnock and smooth flow values:

$$z_0 = 0.11u_*/\nu + \alpha u_*^2/g. \tag{14}$$

Many current bulk algorithms, such as Fairall *et al.* [65, 66] and Zeng *et al.* [67], use eqn (14) for momentum roughness length. These algorithms, as well as Bourassa *et al.* [68], also take into account the effects of gustiness.

Smooth flow effects become significant for wind speeds under about 4 m/s, and lead to an increasing roughness with decreasing wind for $U < 3$ m/s. Although this is qualitatively consistent with the data in Fig. 1, the smooth flow effect is not sufficient to account for the large observed increase in drag at low winds. At $U = 0.5$ m/s, the smooth flow drag coefficient is expected to be 0.0012, significantly lower than the observed range of 0.002–0.005.

Bourassa *et al.* [68] attempted to account for this discrepancy by including the effects of capillary waves. Following Wu [69], the roughness length for capillary waves is given by $z_{0c} = 0.18\sigma/(u_*^2\rho_w)$, where $\sigma$ is the surface tension and $\rho_w$ is the density of water. Including this term in the roughness length leads to a strong increase in $C_{DN}$ as $U$ decreases below 4 m/s, with $C_{DN} \approx 0.002$ for $U \approx 1$ m/s. This is discussed further below.

### 5.2 The effect of sea state: wave age and steepness

Offshore flows near the coast are typically associated with significantly higher drag coefficients than are predicted by eqn (13). This is demonstrated in Fig. 2a, where the WAVES data from the near-shore of Lake Ontario (Drennan *et al.* [45]) are seen to be significantly above eqn (13) on average. In Fig. 2b, all wind sea data from the combined data set are plotted and divided into two groups based on inverse wave age, $u_*/c_p$. It is evident that the younger waves ($u_*/c_p \geq 0.1$, black points) are associated with a higher drag coefficient. This is the so-called wave-age, or fetch, effect.

As shown definitively in the JONSWAP experiment (Hasselmann *et al.* [70]; Komen *et al.* [71]), the wave energy and period increase with fetch $\tilde{X}$ reaching their asymptotic "fully-developed" values for dimensionless fetches ($\tilde{X}g/u_*^2$) of order $10^7$. The dispersive nature of surface waves implies that they propagate more rapidly with increasing wavelength or period. For typical values of wind speed, full development occurs at $\tilde{X} = O(1000\,\text{km})$.

Kitaigorodskii and Volkov [72] included the wave age in their expression for $z_0$, taking into account the speed of the roughness elements relative to the wind. This can be incorporated into the Charnock expression for roughness by assuming the Charnock parameter $\alpha$ to be a function of $c_p/u_*$. This yields an expected drag coefficient relationship of the form $C_{DN} = f(U_{10N}, c_p/u_*)$. Experimental support for this was first presented by Donelan [73]. Subsequent attempts to quantify the relationship were often hampered by several effects. In particular, for many data sets self-correlation between the dimensionless roughness ($z_0/H_s$) and the wave age ($c_p/u_*$) has been a problem. Here the self-correlation arises because $z_0$ is calculated

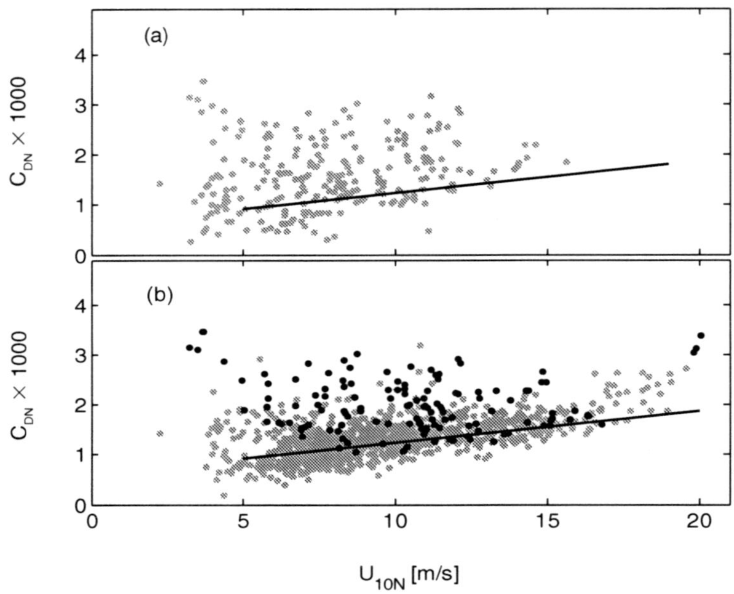


Figure 2: (a) Drag coefficient at 10 m, $C_{\mathrm{DN}} = (u_*/U_{10})^2$ versus 10 m neutral wind speed, $U_{10\mathrm{N}}$, for the WAVES experiment. (b) $C_{\mathrm{DN}}$ versus $U_{10\mathrm{N}}$ for WSD data (eight experiments). Points with $u_*/c_{\mathrm{p}} \geq 0.1$ are black. The Smith [60] bulk relation is shown in both panels.

via eqn (7a) from the friction velocity, and hence both the roughness and the wave age depend on $u_*$.

Recently, Drennan *et al.* [53] using five carefully selected PWS data sets, and an approach designed to minimize spurious correlations, found

$$z_0/H_{\mathrm{s}} = 3.35(c_{\mathrm{p}}/u_*)^{-3.4} \tag{15}$$

or equivalently

$$z_0 g/u_*^2 = 1.7(c_{\mathrm{p}}/u_*)^{-1.7}. \tag{16}$$

Either expression for roughness implies that fetch-limited waves are significantly rougher than their open ocean counterparts (except in duration limited events), and that the fetch-limited momentum transfer (through $C_{\mathrm{DN}}$) is significantly higher. For instance, drag coefficients for young waves with $c_{\mathrm{p}}/u_* = 10$ are 30–50% higher (depending on wind speed) than predicted by eqn (13).

Although these results were derived for PWS data, they are also valid for the less restrictive case of WSD data. In Fig. 3, we plot the dimensionless roughness versus inverse wave age for the combined WSD data. The PWS subset of these data (659 of the 1069 points) is denoted with dots. There is no significant difference between the two sets; eqn (15) is valid for all WSD data, not just PWS. Conversely, eqn (15) is

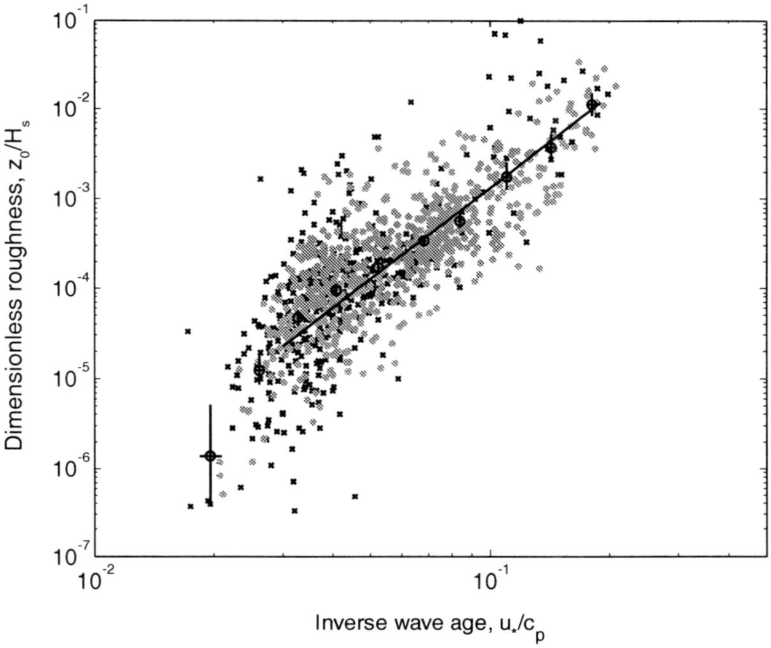


Figure 3: Dimensionless roughness versus inverse wave age for the combined WSD data set. Grey dots denote the PWS subset. The circles show means of data binned into 10 wave-age groups, with 2 standard errors indicated. The line is from Drennan *et al.* [53].

not valid in swell-dominated situations. This is seen in Fig. 4, where the roughness $z_0/H_s$ from the 1395 swell-dominated cases are plotted.

The wave-age effect is strongest within $O(10\,\text{km})$ from the coast, but in stronger winds extends to $O(100\,\text{km})$. During the recent FETCH experiment in the Mediterranean, offshore high wind Mistral flows ($U \sim 16\,\text{m/s}$) produced significantly enhanced stress at the ASIS buoy 50 km offshore (Drennan *et al.* [53]).

Taylor and Yelland [74] have presented an alternative scaling to account for sea state effects. In terms of the wave steepness $H_s/L_p$ they find

$$z_0/H_s = 1200(H_s/L_p)^{4.5}, \qquad (17)$$

where $L_p$ is the peak wavelength. The wave steepness relation was found to be well supported by a variety of data sets, including open ocean mixed-sea conditions, but not strongly forced near-shore (fetch-limited) data.

Drennan *et al.* [75] recently compared, and evaluated, the two sea state formulations. Their conclusions are: (i) in WSD conditions, both eqns (15) and (17) yield significant improvements over standard bulk algorithms; (ii) the wave-age scaling yields the best results in fetch-limited (strongly forced) conditions; (iii) for general

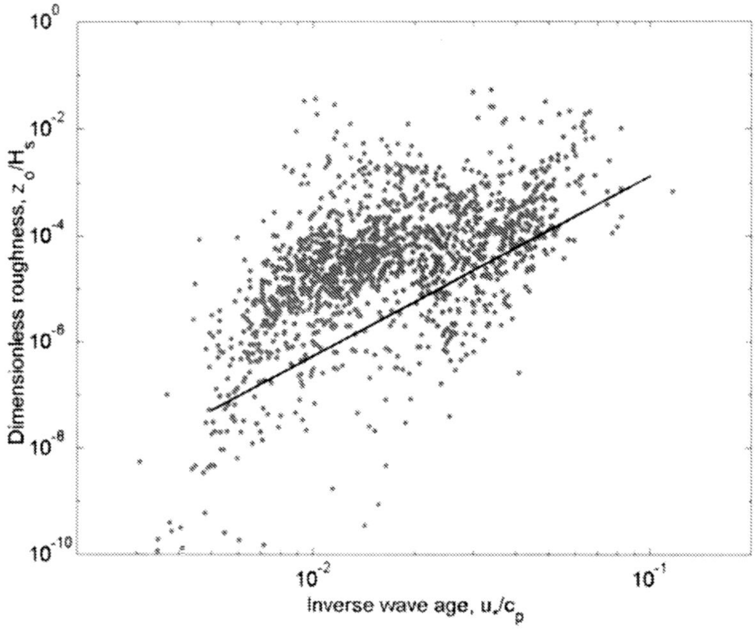


Figure 4: Dimensionless roughness versus inverse wave age for swell-dominated data. The wave-age formula (—) of Drennan *et al.* [53], developed for wind sea conditions, is seen to underpredict most swell data.

mixed seas, the wave steepness is preferred over both wave age and bulk scalings; (iv) in swell-dominated conditions, standard bulk algorithms provide better stress estimates than either sea state scaling.

The application of a sea state formulation requires information (preferably directional) on the wave field. Although many older data sets do not include such parameters, many modern ones do—either from direct or satellite measurements, or via wave models (e.g. Komen *et al.* [71]). To date, few bulk algorithms take sea state effects into account. Exceptions are Bourassa *et al.* [68], Clayson *et al.* [76], and Fairall *et al.* [66]. The first two combine the Smith *et al.* [47] wave-age formulation $z_0 g/u_*^2 = 0.48(c_p/u_*)^{-1}$, similar to eqn (16), with a capillary wave term for low winds. Fairall *et al.* [66] include both wave-age and wave steepness effects as optional inputs.

We emphasize here that the use of a sea state parameterization as part of a general bulk formulation should only be carried out in the appropriate sea state conditions. Wave-age formulations require wind sea parameters, hence are best suited for WSD conditions. At a minimum, the Pierson–Moskowitz criterion $U_{10}/c_p > 0.83$ should be met. For the wave steepness formulation Drennan *et al.* [75] propose a threshold steepness ($H_s/L_p > 0.02$); for lower steepness (most swell-dominated conditions) a simple wind speed dependent $C_{DN}$ yields better predictions than either sea state formulation.

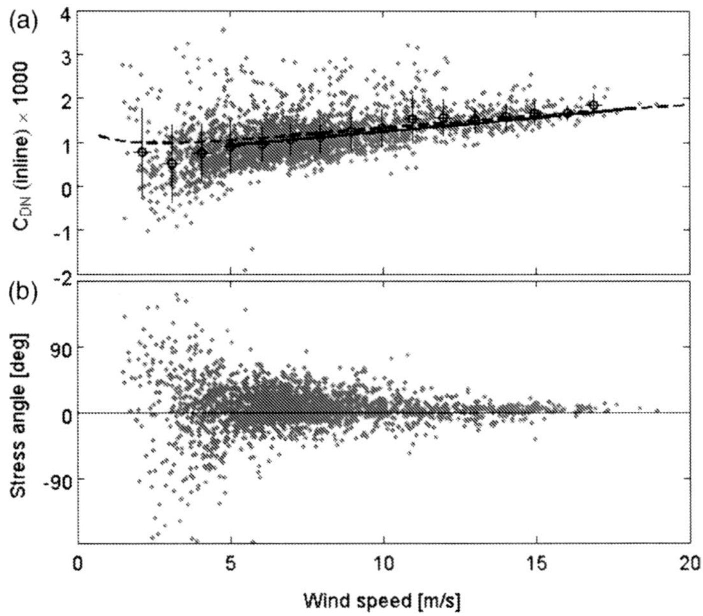


Figure 5: (a) Component of drag coefficient inline with the mean wind, $C_{DX} = -\overline{w'u'}/(U_{10N} - U_0)^2$, versus wind speed $U_{10N}$. The circles represent data averaged in wind speed bins of 1 m/s, showing 1 standard deviation in $U$ and $C_{DX}$. The curves are for $C_{DN}$ via Smith [60] (solid line) and Smith [64], with $z_0 = 0.11\,u_*/\nu + 0.012u_*^2/g$ (dashed line). (b) Stress angle with respect to the mean wind, $a\tan(\overline{w'v'}/\overline{w'u'})$, versus $U_{10N}$.

### 5.3 Sea state effects: swell

The effect of both gustiness and smooth flow would serve to increase the drag coefficient at low winds. Capillary waves have the same effect, assuming they travel in the direction of the mean wind. However, the early measurements of Volkov [16] and Makova [17] indicated significant periods of low wind where the momentum flux was lower than expected. On occasions, momentum flux was even observed to be negative (upward). These periods coincided with the presence of significant swell waves, and led the authors to speculate on a possible influence of swell on the momentum flux. Typical presentations of the drag coefficient, including Fig. 1, hide this effect as they show the magnitude of $C_{DN}$, which is by definition positive. Instead, in Fig. 5 we plot the component of the drag coefficient inline with the mean wind; i.e. $C_{DX} = -\overline{w'u'}/(U_{10N} - U_0)^2$. This plot excludes data from the RASEX and HEXOS experiments where only the stress magnitude is available.

It is evident that many of the low wind/high drag values in Fig. 1 have disappeared, with the data now evenly distributed about the [extended] Smith curve, even at low winds. The dashed line shows $C_{DN}$ based on the Smith [64] roughness (14) which includes a smooth flow correction. The Charnock parameter of 0.012, again calculated from the mean of the data, is lower than that determined earlier due to the use of a single component of stress here, instead of the magnitude. For a similar reason, the agreement with the Smith [60] curve has improved: Smith's drag coefficient was calculated from the component of stress inline with the mean wind. We note here that both definitions for $C_{DN}$ are commonly used in the literature. The inertial dissipation method provides a third estimate, one based on the magnitude of the inline stress component.

Contrary to the situation on land (and the assumptions of the MO theory), the angle of the near-surface stress over the sea does not necessarily coincide with the angle of the wind (Fig. 5b). This was reported by Geernaert [77] who attributed it to stratification effects. Subsequent studies by Geernaert *et al.* [78] and Rieder *et al.* [79] indicate that turning is due to the effect of long waves (swells) on the short waves which form the dominant roughness scales. Grachev *et al.* [80] separated the total stress into components associated with (and traveling in the direction of) the wind waves and swell waves. The total stress was found to be oriented between the wind waves and swell. In some cases when the swell traveled in the same direction as (and faster than) the wind, the total stress was against the wind. In these cases $C_{DX} < 0$, corresponding to upward momentum flux. This phenomenon, seen in the following swell data of Fig. 6, was first noted by Volkov [16] and Makova [17]. The opposite situation with counter-swell (i.e. swell running against the wind) is seen, in Fig. 6, to yield enhanced drag coefficients (Donelan *et al.* [37]). In most following- and counter-swell cases in Fig. 6, the swells are dominant with $E(\text{ws}) \ll E(\text{swell})$.

Over the sea, the momentum flux is given by $\tau = \tau_t + \tau_w + \tau_v$, which are the turbulent, wave-coherent, and viscous stresses, respectively. The viscous stress is significant only in a layer of $O(1\text{ mm})$ above the surface, and is neglected here. The wave-coherent stress $\tau_w$ accounts for the transfer of momentum to the waves. Away from the surface $\tau_w \approx 0$, and the turbulent stress dominates the total stress. This is the classical case of boundary layer flows over land. However, as the surface is approached, there is a region where $\tau_w$ is significant, the so-called "wave boundary layer" (WBL). Since the total stress $\tau$ is constant, the turbulent stress (1a) is consequently reduced in the WBL. This results in the neutral wind speed profiles deviating from their expected logarithmic forms (Stewart [81]). As a result, the MO similarity theory does not apply in the WBL.

The turbulence models of Janssen [82] and Makin and Mastenbroek [83] have predicted the height of the WBL to be of the order of 1 m for PWS conditions. This would imply that the MO theory holds at typical measurement heights of $O(10\text{ m})$ in PWSs, and, indeed, recent experimental evidence (Edson and Fairall [84]; Drennan *et al.* [45]) supports this. In particular, Drennan *et al.* [45] showed that PWS spectra and cospectra representing a wide variety of wind speeds and wave ages ($U_{10N} = 3.5$–$15$ m/s; $U_{10N}/c_p = 1$–$5$), and heights as low as 2 m above the surface, follow the universal curves of Miyake *et al.* [14]. Hence PWS data are expected to follow the MO theory, except perhaps very close to the surface.

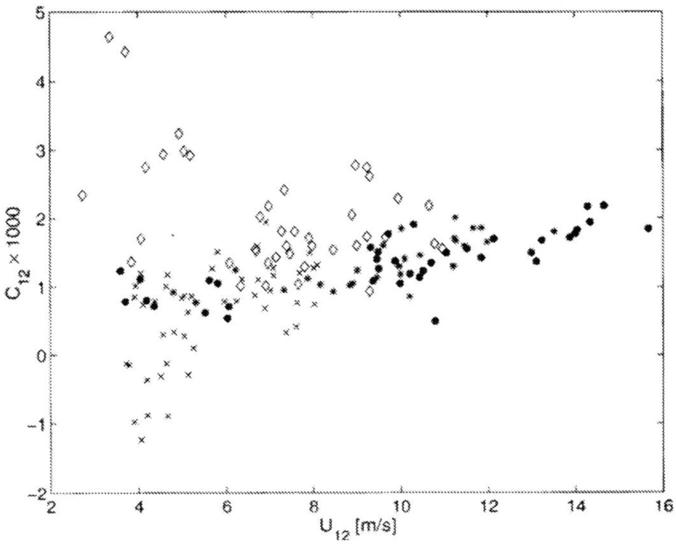


Figure 6: 12 m drag coefficients versus wind speed, showing effect of swell. Data are: PWS (●) and following swell (×) from the WAVES tower in Lake Ontario (Drennan *et al.* [45]), and PWS (∗) and counter swell (◇) from a small ship in the North Atlantic during SWADE (Donelan *et al.* [37]). The two PWS data sets, near full development, are in good agreement. The counter swell data show significantly enhanced drag; the following-swell data significantly reduced drag or even upward momentum transfer. [Reprinted from Drennan *et al.* [45] with kind permission of Kluwer Academic Publishers.]

On the other hand, in the presence of swells the WBL may extend well beyond typical measurement heights. Smedman *et al.* [85, 86] demonstrated that both the wind profiles and the TKE budget are modified when swells are present. Holland *et al.* [18] reported measurements of Donelan showing a reversal of the near-surface logarithmic wind profile ("wave-driven wind") during swells. Drennan *et al.* [45] showed that the universal spectral scaling predicted by the MO theory does not apply in swell-dominated conditions; in contrast, the spectra were supported by the wind sea data. They showed that velocity spectra (Fig. 7) and cospectra with weak–moderate winds and strong swell do not satisfy universal scaling—or MO theory—at typical measurement heights. These findings show that the wave field can significantly modify turbulence characteristics in the marine atmospheric boundary layer, and may explain some of the wide scatter observed in existing drag coefficient relationships.

To date, studies on the effects of swell on momentum flux have been largely qualitative. It is hoped that several recent experiments, along with modeling efforts (e.g. Kudryavtsev and Makin [87]), will yield considerable progress in this field in the next few years.

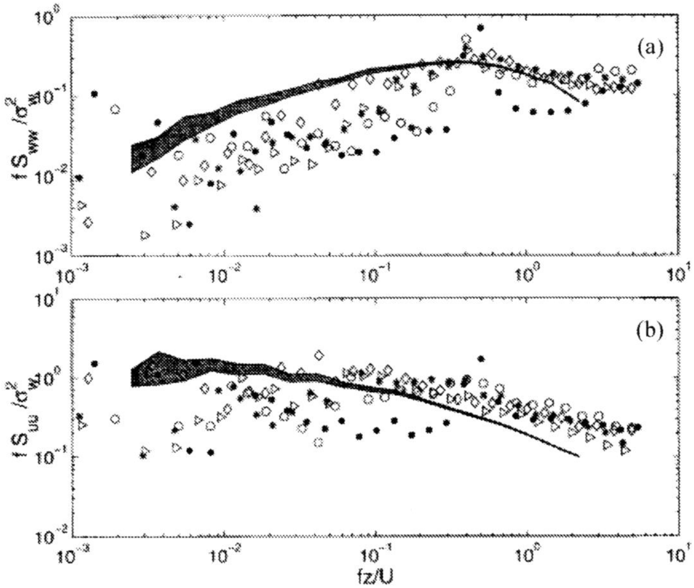


Figure 7: Spectra of vertical (a) and horizontal inline (b) velocity from five runs with strong swell, plotted in the universal scaling of Miyake *et al.* [14]. The data are in the range $U \approx 4$–5 m/s. For the ◇ data, $U/c_p = 0.61$; for the other data, $U/c_p < 0.5$. The shaded areas represent $\pm 1$ standard error about universal curves established from PWS data. [Reprinted from Drennan *et al.* [45] with kind permission of Kluwer Academic Publishers].

## 6 Scalar fluxes

Compared with momentum fluxes, there are relatively few scalar flux data. Although suitable anemometers have been available for decades, reliable fast-response hygrometers have been marketed only for a few years. Fast-response temperature measurements are also limited over the sea due to problems with sea spray and salt accretion on the sensors. An alternate measure of fast-response temperature can be obtained via sonic anemometry. Sonic temperature $t_s$, obtained from the sonic speed of sound (after corrections for cross-flow effects—see Kaimal and Gaynor [88]) is a combination of air temperature and mixing ratio: $t_s = \theta(1+0.52r)$. Hence a sonic anemometer paired with a fast-response hygrometer (or refractometer—see Delahaye *et al.* [89]) will yield the flux of sensible heat, as well as those of humidity and momentum. If a hygrometer is not available, $t_s$ is often used as a surrogate for virtual temperature $t_v = \theta(1+0.61r)$ in calculating the buoyancy flux $F_B$. Sjöblom and Smedman [90] show that the mean error is this approximation is $O(1\%)$.

The HEXOS campaign of 1984–1986 provided what remains the most comprehensive set of sensible and latent heat flux data. The resulting Stanton and Dalton numbers show considerable scatter, but no significant dependence on wind speed for winds in the range 5–20 m/s (DeCosmo *et al.* [48]). Their constant bulk values of $1.1 \times 10^{-3}(\pm 30\%)$ for both $C_{HN}$ and $C_{EN}$ are in the range of values $1 - 1.2 \times 10^{-3}$ given by most other studies (e.g. Large and Pond [31]).

The very different behavior of the scalar and momentum fluxes with wind speed follows from the different physical transfer mechanisms. Form drag over the roughness elements plays an important role in momentum transfer at the air–sea interface, but not in the transfer of scalar quantities. Instead, these latter transfers are controlled by diffusive processes. The well known LKB surface renewal model of Liu, Katsaros, and Businger [91] accounts for these effects, yielding roughness lengths which are functions of the roughness Reynolds number: $z_{0x}u_*/\nu = A_x Re_*^{B_x}$, where $x$ is either $q$ or $\theta$ and $A_x$ and $B_x$ vary with $Re_*$. The resulting expressions predict Stanton and Dalton numbers which decrease somewhat with increasing wind speed. This decrease at high winds is attributed to an increase in sheltering behind roughness elements, which increases the local turbulent renewal times, thereby reducing the transfer rates. The LKB model has been modified for use in some bulk algorithms (COARE-2.5, Fairall *et al.* [65]).

An alternative parameterization for the scalar roughness lengths, that of Brutsaert [92], is also based on surface renewal theory, with the additional assumption that the surface renewal time scale is the same as the Kolmogorov time scale for small eddies (Garratt [3]). The resulting humidity and temperature roughness lengths are $z_{0q} = z_0 \exp(2 - 2.28 Re_*^{1/4})$ and $z_{0t} = z_0 \exp(2 - 2.48 Re_*^{1/4})$, where the coefficients are those of Garratt [3]. The resulting bulk coefficients either tend to a constant value or increase somewhat (Zeng *et al.* [67]) with wind speed. This parameterization has been adopted by several current bulk algorithms, including Zeng *et al.* [67] with $z_{0q} = z_{0t} = z_0 \exp(2.57 - 2.67 Re_*^{1/4})$, and also in a modified form by COARE-3 (Fairall *et al.* [66]).

Figure 8 presents the heat flux data from several of the recent field experiments discussed above. In SWADE, temperature and humidity fluctuations were measured with a fine wire thermocouple and a Lyman-alpha humidiometer, respectively, both housed in a spray flinger (Katsaros *et al.* [51]). During the AGILE campaign, a fine wire dry thermistor and wicked wet-bulb thermistor were used to measure temperature and humidity (Donelan and Drennan [93]). Humidity fluctuations were measured during GASEX-2001 with a LICOR-7500 open path humidiometer.

The Stanton numbers from SWADE and AGILE for $|\Delta\Theta| = |\Theta_z - \Theta_0| > 2°C$ are plotted in Fig. 8a. The threshold is applied taking into account that the measured "surface" $\Theta_0$ is rarely the true skin temperature but instead a bulk temperature at several meters depth. Since the bulk and skin temperatures can differ by $O(1°C)$—see Fairall *et al.* [94]—the threshold reduces the error in $\Delta\Theta$, with the value of 2°C chosen so as to reduce the scatter.

The $C_{HN}$ data show considerable variability, but no significant wind speed dependence. There is, however, a significant difference between the Stanton numbers for stable and unstable stratification, with the stable data exhibiting considerably lower $C_{HN}$ values. This was observed by Large and Pond [31], and Oost *et al.* [95].

Dalton numbers from SWADE, AGILE, and GASEX are plotted in Fig. 8b. Although there is considerable scatter, the three data sets are in agreement, with

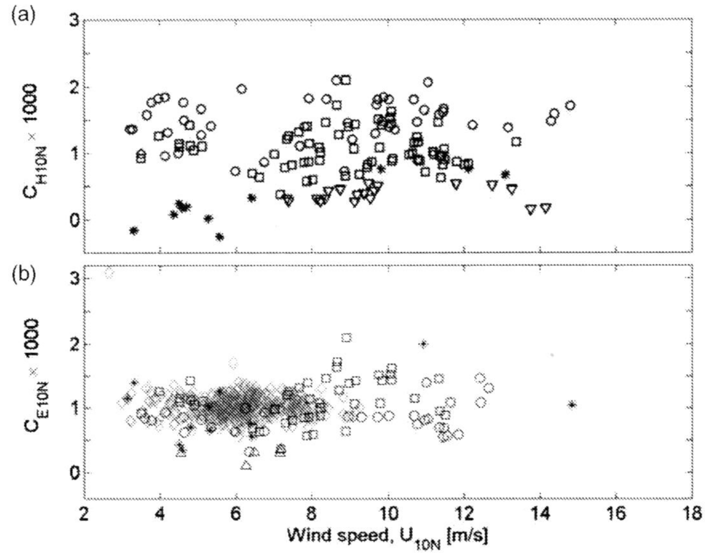


Figure 8: (a) Stanton numbers from two field experiments. SWADE (Katsaros *et al.* [51]): □ (unstable), ∇ (stable); AGILE (Donelan and Drennan [93]): ○ (unstable), ∗ (stable). (b) Dalton numbers from three field campaigns: □, ○, ∗ (as above); AGILE: Δ (condensing); GASEX: ◇ (unstable).

a mean $C_{EN} = 1.02 \pm 0.2 \times 10^{-3}$. The data do not exhibit significant wind speed dependence, nor is there a dependence on stability. However, the few points taken in condensing flows (downward humidity transport) exhibit a significantly lower bulk coefficient.

For several decades, it has been predicted that the presence of sea spray should lead to increased evaporation rates at high winds (e.g. Andreas *et al.* [96]). This claim, which has been the source of several vigorous debates (summarized in Andreas and DeCosmo [97]), remains to be supported by experimental data. Although DeCosmo *et al.* [48] did not find a significant wind speed dependence in the HEXOS $C_{EN}$ values, Andreas and DeCosmo [97] claim a spray enhancement of 10–40% in the same data for $U \sim 15$–18 m/s. However, this enhancement is compared with the values predicted by an LKB-type model which, as pointed out above, decrease at high winds. The net effect is the observed constant Dalton number.

The large scatter in typical Dalton and Stanton number plots (e.g. Fig. 8) continues to hinder efforts to parameterize the bulk coefficients. The data do not allow one to distinguish between simple constant bulk coefficients and theoretical predictions where the roughness is a function of $Re_*$. Likewise, the data alone do not show an enhancement of humidity flux due to sea spray.

Finally, we briefly summarize current progress on $CO_2$ flux parameterization, referring the reader to Fairall *et al.* [5] for further details. As reliable fast-response

$CO_2$ sensors have only been available for a few years, micrometeorological estimates of $CO_2$ flux are relatively rare (McGillis *et al.* [38, 55]; Donelan and Drennan [50]; Jacobs *et al.* [98]). Although these new results have considerable scatter, they agree fairly well with the earlier chemical tracer estimates of Wanninkhof [99] and Watson *et al.* [100]. Nevertheless, there remains considerable uncertainty on the wind speed dependence of the $CO_2$ bulk transfer coefficient, especially at high winds, and on what other factors control the air–sea flux. This remains a subject of active research.

## 7 Discussion

We have summarized the current state of air–sea flux parameterizations and, in particular, how they depend on sea state. Two separate sea state effects were discussed. The "wave-age" effect relates to the increased roughness (momentum transfer) over young, fetch- or duration-limited, waves compared with fully developed ones. Wave age has been included in some recent bulk parameterizations, although additional steps must be taken to ensure that the algorithm is only applied when wind sea parameters are available (from 2D wave spectra, or in WSD conditions).

The swell effect, which remains qualitative at this point, can be significant at low–moderate winds, when strong swells are present. Swell waves can either increase or decrease the drag, depending on the relative direction and amplitude of the swell compared with the wind sea. Bourassa *et al.* [68] attempted to include swell effects in their bulk algorithm by including wave direction in a wave-age term. However, wave-age expressions are developed for wind sea conditions and are not applicable to swell. The inclusion of swell effects in bulk parameterizations awaits a better understanding of how swell influences the air–sea transfers.

Regarding the fluxes of humidity and sensible heat, experimental evidence to date supports constant bulk coefficients, independent of wind speed. The scatter in the data is sufficient that a slight increase (or decrease) in $C_{EN}$ with wind speed cannot be ruled out, but it is not significant. As yet, there is no support for a spray-induced enhancement of $C_{EN}$ above the constant value.

One limitation of all bulk parameterizations is their application in high wind conditions. To date, very few data have been collected at wind speeds above 20 m/s. Hence, the algorithms remain unvalidated in this regime. Although it is common practice to simply extrapolate the present curves to higher winds, this approach is questionable, as characteristics of the sea surface change with increasing wind.

As the wind speed increases the waves become steeper and breaking is more widespread—see for instance Fig. 9, a photograph of the sea surface taken from 470 m during Hurricane Isabel. The laboratory study of Reul *et al.* [101] demonstrates that flow separation increases significantly behind steep breaking waves. Large quantities of spray are also produced, which may effect heat and humidity transfer (as discussed above), and perhaps momentum transfer if the spray density becomes sufficiently high. As the production of both spray and bubbles increases the surface itself is no longer well defined, potentially affecting all transfer processes.

Figure 9: Sea surface between rain bands of category 5 Hurricane Isabel. Photo taken by the author from NOAA–AOC aircraft N43RF on September 14, 2003 during the CBLAST experiment. The altitude is 470 m and the flight level wind speed is 50 m/s.

One topic of considerable recent interest concerns hurricanes. The ongoing ONR-sponsored Coupled Boundary Layer/Air–Sea Transfer (CBLAST) hurricane experiment has a primary focus on increasing our understanding of tropical storm dynamics, and on improving hurricane model accuracy. Hurricanes present a problem for today's bulk parameterizations. Bulk parameterizations predict drag coefficients as increasing roughly linearly with wind speed, but constant Dalton and Stanton numbers. For instance, using the results of Smith [60] at 35 m/s, $C_D \approx 0.003$ and $C_E \approx 0.0012$. The wave-age formulation (15), using wave spectral data from a scanning radar altimeter (SRA) in Hurricane Bonnie, Wright *et al.* [102], predicts drag coefficients roughly 50% higher than Smith [60]—see Fig. 10. Here the predictions for Hurricane Bonnie are estimated only for wave spectra exhibiting a distinct wind sea component.

However, Emanuel [104] has shown that typical hurricane models cannot develop a model storm under the extrapolated predictions of present bulk formulae. Model hurricanes develop only when momentum and enthalpy exchange coefficients are roughly equal. Either the enthalpy flux (a combination of sensible and latent heat fluxes) must increase at higher winds or the drag coefficient must decrease. Powell *et al.* [105] have presented data in support of the latter: their estimates of $u_*$, based on vertical profiles of wind speed from GPS dropsondes, level off when wind speeds reach hurricane levels; even at wind speeds below 30 m/s their drag values are significantly below the Smith [60] predictions. At the same time, Andreas and Emanuel [106] have shown that spray can significantly increase the enthalpy flux if it reenters the sea before completely evaporating.

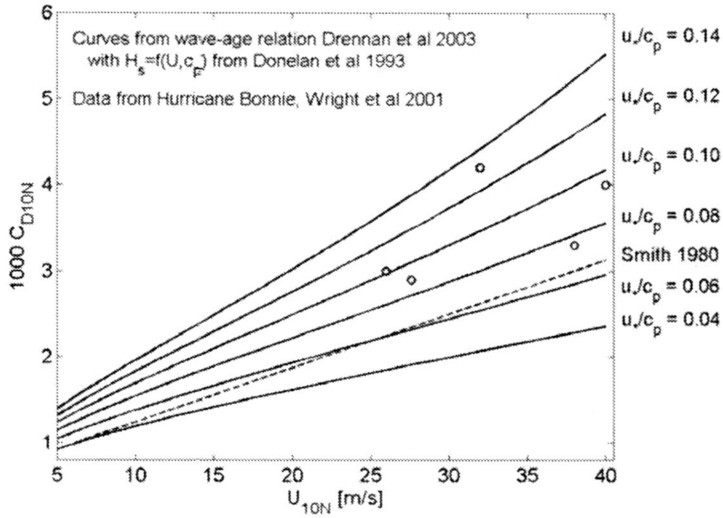


Figure 10: Estimated drag coefficient versus wind speed during Hurricane Bonnie. The estimates are made using SRA wave spectra from Wright *et al.* [102], along with the wave-age relations of Drennan *et al.* [53]. The curves are from [53], with the $H_S$ relation of Donelan *et al.* [103].

Clearly there is a need for measurements of momentum, heat, and humidity fluxes, along with sea state and sea spray, in the high-wind regime. This is one of the goals of CBLAST, and the 2003 campaign has yielded a promising data set. The results should allow for improved flux parameterization at high winds, and ultimately improved coupled ocean–atmosphere models.

## Acknowledgments

Financial support from the US National Science Foundation (grant OCE-9871855) and the US Office of Naval Research (grants N00014-00-1-0594 and N00014-01-F-009) is gratefully acknowledged. Thanks also to Dr. Will Perrie for his helpful comments and suggestions, and to Patricia Archuleta and Maria Calderin for assistance with the typesetting.

## References

[1] Shindell, D.T., Miller, R.L., Schmidt, G.A. & Pandolfo, L., Simulation of recent northern winter climate trends by greenhouse-gas forcing. *Nature*, **399**, pp. 452–455, 1999.

[2] Bao, J.-W., Wilczak, J.M., Choi, J.-K. & Kantha, L.H., Numerical simulations of air–sea interaction under high wind conditions using a coupled

model: a study of hurricane development. *Monthly Weather Review*, **128**, pp. 2190–2210, 2000.

[3] Garratt, J.R., *The Atmospheric Boundary Layer*, Cambridge University Press: Cambridge and New York, 1992.

[4] Webb, E.K., Pearman, G.I. & Leuning, R., Correction of flux measurements for density effects due to heat and water vapour transfer. *Quarterly Journal of the Royal Meteorological Society*, **106**, pp. 85–100, 1980.

[5] Fairall, C.W., Hare, J.E., Edson, J.B. & McGillis, W., Parameterization and micrometeorological measurement of air-sea gas transfer. *Boundary-Layer Meteorology*, **96**, pp. 63–105, 2000.

[6] Monin, A.S. & Obukhov, A.M., Basic laws of turbulent mixing in the ground layer of the atmosphere. *Akad. Nauk. SSSR Geofiz. Inst. Tr.*, **151**, pp. 163–187, 1954.

[7] Busch, N.E., On the mechanics of atmospheric turbulence (Chapter 1). *Workshop on Micro-meteorology*, ed. D.A. Haugen, American Meteorological Society: Boston, pp. 1–65, 1973.

[8] Donelan, M.A., Air–sea interaction. *The Sea: Ocean Engineering Science*, 9, eds B. LeMéhauté D. Hanes, John Wiley and Sons, Inc.: New York, pp. 239–292, 1990.

[9] Högström, U., Nondimensional wind and temperature profiles in the atmospheric surface layer: a re-evaluation. *Boundary-Layer Meteorology*, **42**, pp. 55–78, 1988.

[10] Zilitinkevich, S.S. & Chalikov, D.V., Determining the universal wind velocity and temperature profiles in the atmospheric boundary layer. *Izvestiya, Atmospheric and Oceanic Physics*, **4**, pp. 165–170, 1968.

[11] Dyer, A.J. & Hicks, B.B., Flux-gradient relationships in the constant flux layer. *Quarterly Journal of the Royal Meteorological Society*, **96**, pp. 715–721, 1970.

[12] Businger, J.A., Wyngaard, J.C., Izumi, Y.K. & Bradley, E.F., Flux-profile relationships in the atmosphere surface layer. *Journal of the Atmospheric Sciences*, **28**, pp. 181–189, 1971.

[13] Högström, U., Review of some basic characteristics of the atmospheric surface layer. *Boundary-Layer Meteorology*, **78**, pp. 215–246, 1996.

[14] Miyake, M., Stewart, W. & Burling, R.W. Spectra and cospectra of turbulence over water. *Quarterly Journal of the Royal Meteorological Society*, **96**, pp. 138–143, 1970.

[15] Pond, S., Phelps, G.T., Paquin, J.E., McBean, G. & Stewart, R.W. Measurements of the turbulent fluxes of momentum, moisture and sensible heat over the ocean. *Journal of the Atmospheric Sciences*, **28**, pp. 901–917, 1971.

[16] Volkov, Y.A., Turbulent flux of momentum and heat in the atmospheric surface layer over a disturbed surface. *Izvestiya, Atmospheric and Oceanic Physics*, **6**, pp. 770–774, 1970.

[17] Makova, V.I., Features of the dynamics of turbulence in the marine atmospheric surface layer at various stages in the development of waves. *Izvestiya, Atmospheric and Oceanic Physics*, **11**, pp. 297–307, 1975.

[18] Holland, J.Z., Chen, W., Almazan, J.A. & Elder, F.C., Atmospheric boundary layer. *IFYGL—The International Field Year for the Great Lakes*, eds E.J. Aubert & T.L. Richards, NOAA: Ann Arbor, pp. 109–167, 1981.

[19] Paulson, C.A., Leavitt, E. & Fleagle, R.G. Air-sea transfer of momentum, heat and water determined from profile measurements during BOMEX. *Journal of Physical Oceanography*, **2**, pp. 487–497, 1972.

[20] Davidson, K.L., Observational results on the influence of stability and wind-wave coupling on momentum transfer and turbulent fluctuations over ocean waves. *Boundary-Layer Meteorology*, **6**, pp. 305–331, 1974.

[21] Kaimal, J.C., Wyngaard, J.C., Izumi, Y. & Coté, O.R., Spectral characteristics of surface-layer turbulence. *Quarterly Journal of the Royal Meteorological Society*, **98**, pp. 563–589, 1972.

[22] Mahrt, L., Vickers, D., Howell, J., Højstrup, J., Wilczak, J.M., Edson, J. & Hare, J., Sea surface drag coefficients in the Risø Air Sea Experiment. *Journal of Geophysical Research*, **101**, pp. 14327–14335, 1996.

[23] Axford, D.N., On the accuracy of wind measurements using an inertial platform in an aircraft, and an example of a measurement of the vertical mesostructure of the atmosphere. *Journal of Applied Meteorology*, **7**, pp. 645–666, 1968.

[24] Anctil, F., Donelan, M.A., Drennan, W.M. & Graber, H.C., Eddy correlation measurements of air-sea fluxes from a Discus buoy. *Journal of Atmospheric and Oceanic Technology*, **11**, pp. 1144–1150, 1994.

[25] Edson, J.B., Hinton, A.A., Prada, K.E., Hare, J.E. & Fairall, C.W., Direct covariance flux estimates from mobile platforms at sea. *Journal of Atmospheric and Oceanic Technology*, **15**, pp. 547–562, 1998.

[26] Wieringa, J., A revaluation of the Kansas mast influence on measurements of stress and cup anemometer overspeeding. *Boundary-Layer Meteorology*, **18**, pp. 411–430, 1980.

[27] Wyngaard, J.C., The effects of probe-induced flow distortion on atmospheric turbulence measurements. *Journal of Applied Meteorology*, **20**, pp. 784–794, 1981.

[28] Yelland, M.J., Moat, B.I., Taylor, P.K., Pascal, R.W., Hutchings, J. & Cornell, V.C., Wind stress measurements from the open ocean corrected for airflow distortion by the ship. *Journal of Physical Oceanography*, **28**, pp. 1511–1526, 1998.

[29] Dupuis, H., Guérin, C., Hauser, D., Weill, A., Nacass, P., Drennan, W.M., Cloché, S. & Graber, H.C., Impact of flow distortion corrections on turbulent fluxes estimated by the inertial dissipation method during the FETCH experiment on R/V L'Atalante. *Journal of Geophysical Research*, **108**(C3), 8064, doi:10.1029/2001JC001075, 2003.

[30] Pedreros, R., Dardier, G., Dupuis, H., Graber, H.C., Drennan, W.M., Weill, A., Guérin, C. & Nacass, P., Momentum and heat fluxes via the eddy correlation method on the R/V L'Atalante and an ASIS buoy. *Journal of Geophysical Research*, **108**(C11), 3339, doi:10.1029/2002JC001449, 2003.

[31] Large, W.G. & Pond, S., Sensible and latent heat flux measurements over the ocean. *Journal of Physical Oceanography*, **12**, pp. 464–482, 1982.

[32] Large, W.G. & Pond, S., Open ocean momentum flux measurements in moderate to strong winds. *Journal of Physical Oceanography*, **11**, pp. 324–336, 1981.

[33] Dupuis, H., Taylor, P.K., Weill, A. & Katsaros, K., Inertial dissipation method applied to derive turbulent fluxes over the ocean during the Surface of the Ocean, Fluxes and Interactions with the Atmosphere/Atlantic Stratocumulus Transition Experiment (SOFIA/ASTEX) and Structure des Echanges
Mer-Atmosphere, Proprietes des Heterogeneites Oceaniques: Recherche Experimentale (SEMAPHORE) experiments with low to moderate wind speeds. *Journal of Geophysical Research*, **102**, pp. 21115–21129, 1997.

[34] Sjöblom, A. & Smedman, A.S., Vertical structure in the marine atmospheric boundary layer and its implication for the inertial dissipation method. *Boundary-Layer Meteorology*, **109**, pp. 1–25, 2003.

[35] Janssen, P.A.E.M., On the effect of ocean waves on the kinetic energy balance and consequences for the inertial dissipation method. *Journal of Physical Oceanography*, **29**, pp. 530–534, 1999.

[36] Taylor, P.K. & Yelland, M.J., Comments on "On the effect of ocean waves on the kinetic energy balance and consequences for the inertial dissipation method". *Journal of Physical Oceanography*, **31**, pp. 2532–2536, 2001.

[37] Donelan, M.A., Drennan, W.M. & Katsaros, K.B., The air-sea momentum flux in conditions of mixed wind sea and swell. *Journal of Physical Oceanography*, **27**, pp. 2087–2099, 1997.

[38] McGillis, W.R., Edson, J.B., Ware, J.D., Dacey, J.W.H, Hare, J.E., Fairall, C.W. & Wanninkhof, R., Carbon dioxide flux techniques performed during GasEx-98. *Marine Chemistry*, **75**, pp. 267–280, 2001.

[39] Brunke, M.A., Fairall, C.W., Zeng, X., Eymard, L. & Curry, J.A., Which bulk aerodynamic algorithms are least problematic in computing ocean surface turbulent fluxes? *Journal of Climate*, **16**, pp. 619–635, 2003.

[40] Charnock, H., Wind stress on a water surface. *Quarterly Journal of the Royal Meteorological Society*, **81**, pp. 639–640, 1955.

[41] Kraus E.B. & Businger, J.A., *Atmosphere-Ocean Interaction.* Oxford University Press: New York, 1994.

[42] Geernaert, G.L., Bulk parameterization for the wind stress and heat fluxes. *Surface Waves and Fluxes*, eds G.L. Geernaert & W.J. Plant, Kluwer: Dordrecht, pp. 91–172, 1990.

[43] Sreenivasan, K.R., Chambers, A.J. & Antonia, R.J., Accuracy of moments of velocity and scalar fluctuations in the atmospheric surface layer. *Boundary-Layer Meteorology*, **14**, pp. 341–359, 1978.

[44] Donelan, M.A., Madsen, N., Kahma, K., Tsanis, I.K. & Drennan, W.M., Apparatus for atmospheric boundary layer measurements over waves. *Journal of Atmospheric and Oceanic Technology*, **16**, pp. 1172–1182, 1999.

[45] Drennan, W.M., Kahma, K.K. & Donelan, M.A., On momentum flux and velocity spectra over waves. *Boundary-Layer Meteorology*, **92**, pp. 489–515, 1999.

[46] Johnson, H.K., Højstrup, J., Vested, H.J. & Larsen, S.E., On the dependence of sea surface roughness on wind waves. *Journal of Physical Oceanography*, **28**, pp. 1702–1716, 1998.

[47] Smith, S.D., Anderson, R.J., Oost, W.A., Kraan, C., Maat, N., deCosmo, J., Katsaros, K.B., Davidson, K.L., Bumke, K., Hasse, L. & Chadwick, H.M., Sea surface wind stress and drag coefficients: the HEXOS results. *Boundary-Layer Meteorology*, **60**, pp. 109–142, 1992.

[48] DeCosmo, J., Katsaros, K.B., Smith, S.D., Anderson, R.J., Oost, W.A., Bumke, K. & Chadwick, H., Air-sea exchange of water and sensible heat: The Humidity Exchange Over the Sea (HEXOS) results. *Journal of Geophysical Research*, **101**, pp. 12001–12016, 1996.

[49] Janssen, J.A.M., Does wind stress depend on sea state or not? A statistical analysis of HEXMAX data. *Boundary-Layer Meteorology*, **83**, pp. 479–503, 1997.

[50] Donelan, M.A. & Drennan, W.M., Direct field measurements of the flux of carbon dioxide. *Air-Water Gas Transfer*, eds B. Jähne & E.C. Monahan, Aeon-Verlag: Hanau, pp. 677–683, 1995.

[51] Katsaros, K.B., Donelan, M.A. & Drennan, W.M., Flux measurements from a Swath ship in SWADE. *Journal of Marine Systems*, **4**, pp. 117–132, 1993.

[52] Graber, H.C., Terray, E.A., Donelan, M.A., Drennan, W.M., Van Leer, J. & Peters, D.B., ASIS—A new air-sea interaction spar buoy: design and performance at sea. *Journal of Atmospheric and Oceanic Technology*, **17**, pp. 708–720, 2000.

[53] Drennan, W.M., Graber, H.C., Hauser, D. & Quentin, C., On the wave age dependence of wind stress over pure wind seas. *Journal of Geophysical Research*, **108**(C3), 8062, doi:10.1029/2000JC000715, 2003.

[54] Dhanak, M.R., An, P.E. & Holappa, K., An AUV survey in the littoral zone: small-scale subsurface variability accompanying synoptic observations of surface currents. *IEEE Journal of Oceanic Engineering*, **26**, pp. 752–768, 2003.

[55] McGillis, W.G., Edson, J.B., Zappa, C.J., Ware, J.D., McKenna, S.P., Terray, E.A., Hare, J.E., Fairall, C.W., Drennan, W., Donelan, M., DeGrandpre, M.D., Wanninkhof, R. & Feely, R.A., Air-sea $CO_2$ exchange in the Equatorial Pacific. *Journal of Geophysical Research*, **109**, C08S02, doi:10.1029/2003JC002256, 2004.

[56] Drennan, W.M., Donelan, M.A., Madsen, N., Katsaros, K.B., Terray, E.A. & Flagg, C.N., Directional wave spectra from a Swath ship at sea. *Journal of Atmospheric and Oceanic Technology*, **11**, pp. 1109–1116, 1994.

[57] Pettersson, H., Graber, H.C., Hauser, D., Quentin, C., Kahma, K.K., Drennan, W.M. & Donelan, M.A., Directional wave measurements from three wave sensors during the FETCH experiment. *Journal of Geophysical Research*, **108**(C3), 8061, doi:10.1029/2001JC001164, 2003.

[58] Gerling, T.W., Partitioning sequences and arrays of directional ocean wave spectra into component wave systems. *Journal of Atmospheric and Oceanic Technology*, **9**, pp. 444–458, 1992.

[59] Donelan, M.A., Hamilton, J. & Hui, W.H., Directional spectra of wind generated waves. *Philosophical Transactions of the Royal Society of London*, **A315**, pp. 509–562, 1985.

[60] Smith, S.D., Wind stress and heat flux over the ocean in gale force winds. *Journal of Physical Oceanography*, **10**, pp. 709–726, 1980.

[61] Businger, J.A., A note on free convection. *Boundary-Layer Meteorology*, **4**, pp. 323–326, 1973.

[62] Godfrey, J.S. & Beljaars, A.C.M., On the turbulent fluxes of buoyancy, heat and moisture at the air-sea interface at low wind speeds. *Journal of Geophysical Research*, **96**, pp. 22043–22048, 1991.

[63] Grachev, A.A. & Fairall, C.W., Dependence of the Monin-Obukhov stability parameter on the bulk Richardson number over the ocean. *Journal of Applied Meteorology*, **36**, pp. 406–415, 1997.

[64] Smith, S.D., Coefficients for sea surface wind stress, heat flux and wind profiles as a function of wind speed and temperature. *Journal of Geophysical Research*, **93**, pp. 15467–15472, 1988.

[65] Fairall, C.W., Bradley, E.F., Rogers, D.P., Edson, J.B. & Young, G.S., Bulk parameterization of air-sea fluxes for TOGA COARE. *Journal of Geophysical Research*, **101**, pp. 3747–3764, 1996.

[66] Fairall, C.W., Bradley, E.F., Hare, J.E., Grachev, A.A. & Edson, J.B., Bulk parameterization of air-sea fluxes: updates and verification for the COARE algorithm. *Journal of Climate*, **16**, pp. 571–591, 2003.

[67] Zeng, X., Zhao, M. & Dickinson, R.E., Intercomparison of bulk aerodynamic algorithms for the computation of sea surface fluxes using TOGA COARE and TAO data. *Journal of Climate*, **11**, pp. 2628–2644, 1998.

[68] Bourassa, M.A., Vincent, D.G. & Wood, W.L., A flux parameterization including the effects of capillary waves and sea state. *Journal of the Atmospheric Sciences*, **56**, pp. 1123–1139, 1999.

[69] Wu, J., Laboratory studies of wind-wave interaction. *Journal of Fluid Mechanics*, **34**, pp. 91–112, 1968.

[70] Hasselmann, K., Barnett, T.P., Bouws, E., Carlson, H., Cartwright, D.E., Enke, K., Ewing, J.A., Gienapp, H., Hasselmann, D.E., Kruseman, P., Meerburg, A., Müller, P., Olbers, D.J., Richter, K., Sell, W. & Walden, H., Measurements of wind-wave growth and swell decay during the Joint North Sea Wave Project (JONSWAP). *Deutsche Hydrographische Zeitung*, **A8**(Suppl. 12), 95pp., 1973.

[71] Komen, G.J., Cavaleri, L., Donelan, M., Hasselmann, K., Hasselmann, S. & Janssen, P.A.E.M., *Dynamics and Modelling of Ocean Waves*, Cambridge University Press: Cambridge, 1994.

[72] Kitaigorodskii, S.A. & Volkov, Y.A., On the roughness parameter of the sea surface and the calculation of momentum flux in the near-water layer of the

atmosphere. *Izvestiya, Atmospheric and Oceanic Physics*, **1**, pp. 973–988, 1965.

[73] Donelan, M.A., The dependence of the aerodynamic drag coefficient on wave parameters. *First International Conference on Meteorology and Air-Sea Interaction of the Coastal Zone*, American Meteorological Society: Boston, pp. 381–387, 1982.

[74] Taylor, P.K. & Yelland, M.J., The dependence of sea surface roughness on the height and steepness of the waves. *Journal of Physical Oceanography*, **31**, pp. 572–590, 2000.

[75] Drennan, W.M., Taylor, P.K. & Yelland, M.J., Parameterizing the sea surface roughness. *Journal of Physical Oceanography*, **35**, pp. 835–848, 2005.

[76] Clayson, C.A., Fairall C.W. & Curry, J.A., Evaluation of turbulent fluxes at the ocean surface using surface renewal theory. *Journal of Geophysical Research*, **101**, pp. 28503–28513, 1996.

[77] Geernaert, G.L., Measurements of the angle between the wind vector and wind stress vector in the surface layer over the North Sea. *Journal of Geophysical Research*, **93**, pp. 8215–8220, 1988.

[78] Geernaert, G.L., Hansen, F., Courtney, M. & Herbers, T., Directional attributes of the ocean surface wind stress vector. *Journal of Geophysical Research*, **98**, pp. 16571–16582, 1993.

[79] Rieder, K.F., Smith, J.A. & Weller, R.A., Observed directional characteristics of the wind, wind stress and surface waves on the open ocean. *Journal of Geophysical Research*, **99**, pp. 22589–22596, 1994.

[80] Grachev, A.A., Fairall, C.W., Hare, J.E., Edson, J.B. & Miller, S.D., Wind stress vector over ocean waves. *Journal of Physical Oceanography*, **33**, pp. 2408–2429, 2003.

[81] Stewart, R.W., The wave drag of wind over water. *Journal of Fluid Mechanics*, **10**, pp. 189–194, 1961.

[82] Janssen, P.A.E.M., Wave-induced stress and the drag of air flow over sea waves. *Journal of Physical Oceanography*, **19**, pp. 745–754, 1989.

[83] Makin, V.K. & Mastenbroek, C., Impact of waves on air-sea exchange of sensible heat and momentum. *Boundary-Layer Meteorology*, **79**, pp. 279–300, 1996.

[84] Edson, J.B. & Fairall, C.W., Similarity relationships in the marine atmospheric surface layer for terms in the TKE and scalar variance budgets. *Journal of the Atmospheric Sciences*, **55**, pp. 2311–2328, 1998.

[85] Smedman, A.S., Tjernström, M. & Högström, U., The near-neutral marine atmospheric boundary layer with no surface shearing stress: a case study. *Journal of the Atmospheric Sciences*, **51**, pp. 3399–3411, 1994.

[86] Smedman, A.S., Högström, U., Bergström, H., Rutgersson, A., Kahma, K.K. & Pettersson, H., A case study of air-sea interaction during swell conditions. *Journal of Geophysical Research*, **104**, pp. 25833–25852, 1999.

[87] Kudryavtsev, V.N. & Makin, V.K., Coupled dynamics of short waves and the airflow over long surface waves. *Journal of Geophysical Research*, **107**(C12), 3209, doi:10.1029/2001JC001251, 2002.

[88] Kaimal, J.C. & Gaynor, J.E., Another look at sonic anemometry. *Boundary-Layer Meteorology*, **56**, pp. 401–410, 1991.

[89] Delahaye, J.Y., Guérin, C., Vinson, J.P., Dupuis, H., Weill, A., Branger, H., Eymard, L., Lavergnat, J. & Lachaud, G., A new shipborne microwave refractometer for estimating the evaporation flux at the sea surface. *Journal of Atmospheric and Oceanic Technology*, **18**, pp. 459–475, 2001.

[90] Sjöblom, A. & Smedman, A.S., The turbulent kinetic energy budget in the marine atmospheric surface layer. *Journal of Geophysical Research*, **107**(C10), doi:10.1029/2001JC001016, 2002.

[91] Liu, W.T., Katsaros, K.B. & Businger, J.A., Bulk parameterization of air-sea exchanges of heat and water vapor including the molecular constraints at the interface. *Journal of the Atmospheric Sciences*, **36**, pp. 1722–1735, 1979.

[92] Brutsaert, W., *Evaporation into the Atmosphere*, Reidel: Dordrecht and Boston, 1982.

[93] Donelan M.A. & Drennan, W.M., The effect of surface roughness on the fluxes of $CO_2$ and $H_2O$, manuscript in preparation, 2004.

[94] Fairall, C.W., Bradley, E.F., Godfrey, J.S., Wick, G.A., Edson, J.B. & Young, G.S., Cool-skin and warm-layer effects on sea surface temperature. *Journal of Geophysical Research*, **101**, pp. 1295–1308, 1996.

[95] Oost, W.A., Jacobs, C.M.J. & van Oort, C., Stability effects on heat and moisture fluxes at sea. *Boundary-Layer Meteorology*, **95**, pp. 271–302, 2000.

[96] Andreas, E.L., Edson, J.B., Monahan, E.C., Rouault, M.P. & Smith, S.D., The spray contribution to net evaporation from the sea: a review of recent progress. *Boundary-Layer Meteorology*, **72**, pp. 3–52, 1995.

[97] Andreas, E.L. & DeCosmo, J., The signature of sea spray in the HEXOS turbulent heat flux data. *Boundary-Layer Meteorology*, **103**, pp. 303–333, 2002.

[98] Jacobs C.M.J., Kohsiek, W. & Oost, W.A., Air-sea fluxes and transfer velocity of $CO_2$ over the North Sea: results from ASGAMAGE. *Tellus B*, **51**, pp. 629–641, 1999.

[99] Wanninkhof, R., Relationship between wind speed and gas exchange over the ocean. *Journal of Geophysical Research*, **97**, pp. 7373–7382, 1992.

[100] Watson, A., Upstill-Goddard, R. & Liss, P.S., Air-sea exchange in rough and stormy seas, measured by a dual tracer technique. *Nature*, **349**, pp. 145–147, 1991.

[101] Reul, N., Branger, H. & Giovanangeli, J.P., Air flow separation over unsteady breaking waves. *Physics of Fluids*, **11**, pp. 1959–1961, 1999.

[102] Wright, C.W., Walsh, E.J., Vandemark, D., Krabill, W.B., Garcia, A.W., Houston, S.H., Powell, M.D., Black, P.G. & Marks, F.D., Hurricane directional wave spectrum spatial variation in the open ocean. *Journal of Physical Oceanography*, **31**, pp. 2472–2488, 2001.

[103] Donelan, M.A., Dobson, F.W., Smith S.D. & Anderson, R.J., On the dependence of sea surface roughness on wave development. *Journal of Physical Oceanography*, **23**, pp. 2143–2149, 1993.
[104] Emanuel, K.A., Thermodynamic control of hurricane intensity. *Nature*, **401**, pp. 665–669, 1999.
[105] Powell, M.D., Vickery, P.J. & Reinhold, T.A., Reduced drag coefficient for high wind speeds in tropical cyclones. *Nature*, **422**, pp. 279–283, 2003.
[106] Andreas, E.L. & Emanuel, K.A., Effects of sea spray on tropical cyclone intensity. *Journal of the Atmospheric Sciences*, **58**, pp. 3741–3751, 2001.

# CHAPTER 2

# Satellite-based observations of surface turbulent stress during severe weather

M.A. Bourassa
*Center for Ocean-Atmospheric Prediction Studies and Department of Meteorology, Florida State University, USA.*

## Abstract

Vector surface turbulent stresses are estimated from vector winds determined from space-borne scatterometers. Scatterometer vector winds differ subtly from *in situ* observations of vector winds. These differences are exaggerated for severe storm conditions, which is advantageous for determining surface stress. A physically based model for surface stress is developed to link surface turbulent stress to near-surface winds, and to explain the observed qualitative differences between *in situ* and scatterometer winds. The improvements in this model compared with previous formulations are in the near-surface boundary conditions on wind speed and a vertical offset (i.e., displacement height) of the log-wind profile, due to wave modification of the surface. The boundary condition for wind speed implies that there must be a vertical offset, which can be used to derive the displacement height. This formulation represents a first estimate of the dependence of displacement height on wave characteristics. One advantage of this approach is that it removes sea state dependency from Charnock's constant. Strengths and weaknesses are discussed for this model and for its application in determining scatterometer-based vector surface turbulent stress.

## 1 Introduction

The environment of severe marine weather is harsh: *in situ* and satellite observations of surface turbulent stresses are extremely difficult to acquire under such conditions. Even for fair weather conditions, the spatial density of *in situ* observations is insufficient to examine the spatial and temporal evolution of mesoscale atmospheric systems. Satellite remote sensing provides an alternative that is capable of capturing the spatial structure of mesoscale atmospheric systems, although the temporal sampling is currently less than desired (Schlax *et al.* [1]). The lack of *in situ* stress observations is so severe that satellites are calibrated to measure winds rather than stresses, even though, since the planning stages of the satellite era, it was believed that satellite-borne instruments responded more to stress than to wind (O'Brien *et al.* [2]). Moreover, the frequency of occurrence of surface (10 m) wind speeds greater than 16 m $s^{-1}$ is so rare that calibration of satellite instruments is suspect for wind speeds over about 20 m $s^{-1}$. A further difficulty is that the calculation of surface stress, which is crucial for ocean forcing and momentum budgets on both sides of the air/sea interface, is complicated by the conversion of surface winds to surface turbulent stresses. For very large wind speeds, such as those found in hurricanes, there are several recent indications that the rate of increase in stress with increasing wind speed drops (Powell *et al.* [3]). Moreover, parameterizations of surface turbulent stress, which are calibrated for much lower wind speeds, tend to predict stresses that are too large when extrapolated to such extreme conditions. When such parameterizations are used in modeling strong hurricanes, the hurricanes weaken too fast due to excessive surface stress (Chen [4]). Satellite observations of winds will be discussed in Section 2, and the conversion of wind speed to stress will be addressed in Section 2.3.

Surface stress over water is primarily dependent on the vertical profile of wind speed, which is mainly dependent on the wind speed differences between the sea surface and a known height. Secondary dependencies are the stratification of the atmosphere (atmospheric stability) and sea state (i.e., characteristics of the surface wave field). In severe weather, there can be great variations in sea state, both in the magnitude of the waves and in the direction of wave propagation relative to the wind direction. Physical mechanisms have been proposed to account for the dependence of stress on sea state in many recent studies (Kusaba and Masuda [5]; Geernaert [6]; Toba *et al.* [7]; Perrie and Toulany [8]; Maat *et al.* [9]; Smith *et al.* [10]; Yelland *et al.* [11]; Bourassa *et al.* [12]; Bourassa [13]); however, an empirical formulation by Taylor and Yelland [14] based on a much wider range of wind speeds and nondirectional sea states has been shown to provide a better match to mean observations that cover a wide range of conditions. Taylor and Yelland [14] acknowledge that while their model is a good match to the mean wind speed dependence, it leaves a great deal of variability unexplained. A physically based model that has a similar wind speed dependent mean and accounts for much more variability is discussed in Section 4.

Differences between satellite and *in situ* wind observations have been used to infer the relative importance of several wave characteristics by Quilfen *et al.* [15]. These results are expressed in terms of correlations among variables, which do not explain the physical mechanisms through which these wave characteristics modify surface fluxes. These wind observation differences also suggest that sea state parameterizations should be used to convert the *in situ* and/or satellite wind speeds to stresses. A physical mechanism that utilizes these insights was developed in the flux model of Bourassa [13] and found to be largely consistent with the results of Taylor and Yelland's [14] empirical relation. One advantage of this physically based mechanism is that it also considers directional sea state (i.e., the wind direction relative to the directions of wave propagation). This additional consideration makes the flux model consistent with observations over a wider range of conditions. Concomitant nondirectional impacts on surface turbulent stress, sensible heat, and latent heat fluxes are also discussed. These impacts can be substantial in severe weather.

## 2 Satellite observations of surface winds

In the late 1990s, microwave scatterometry finally caught up to other radiometric instruments of the SeaSat satellite: altimeters (ocean height, wave height), radiometers (temperatures, humidity, and rain), and scatterometers (wind speed and direction) were all designed to provide previously unattainable quantity and quality of data regarding variability of the ocean and the adjoining atmospheric boundary layer (Katsaros and Brown [16]). For example, the scatterometers on the European Remote Sensing Satellite Systems (ERS-1 and ERS-2) provided the first scatterometer data (starting in 1991) that could be used for climatological studies. However, operational constraints prevented continuous scatterometer observations while over water, which was where it was usually operating.

The Japanese satellite, ADEOS, was launched in August 1996 with the first dedicated microwave scatterometer since SeaSat: the NASA Scatterometer (NSCAT). This scatterometer determined wind speed and direction over 90% of the ice-free global water surface every 2 days, with a 25-km in-swath grid spacing. It functioned until a catastrophic failure of the satellite platform occurred on June 29, 1997. Despite this loss, the unprecedented coverage and resolution of global wind data resulted in profound impacts on oceanographic and meteorological applications.

The unprecedented accuracy and coverage of NSCAT winds led to the rapid deployment of a new type of scatterometer (SeaWinds) to attempt to resume the NSCAT-like observations. SeaWinds instruments were deployed on QuikSCAT (July 24, 1999 to date) and ADEOS-2 (April 10–October 22, 2003). SeaWinds scatterometers have approximately double the coverage of NSCAT, covering >90% of the world's oceans everyday. The NSCAT and SeaWinds periods may be the only times to date when ocean modelers could not reasonably argue that errors in ocean model outputs were due mainly to shortcomings in wind observations. The short period when two SeaWinds scatterometers were operational provided approximately six-hourly coverage over most of the oceans.

### 2.1 Physics of scatterometry

Scatterometers are unique among satellite remote sensors in their ability to accurately determine the wind speed and direction. Scatterometer microwaves are Bragg scattered by short water waves, which respond quickly to changes in winds. This backscatter (the fraction of transmitted energy that returns to the satellite) is dependent on wind speed and wind direction. The wind direction is found by determining the angle that is most likely to match that of the observed backscatters, using a digital filtering technique (Naderi *et al.* [17]) to rapidly sample small regions from multiple angles over a small period of time. There are substantial design differences among ERS scatterometers, NSCAT, and SeaWinds. For example, the ERS backscatter is spatially smoothed, thereby reducing the resolution to ~70 km (Freilich and Long [18]). The ERS satellites used three fixed antennas to scan the surface on one side of the satellite, whereas NSCAT had three fixed antennas on each side, allowing swaths on each side of the satellite track to be sampled by fore-, mid-, and aft-beams. NSCAT wind speeds and directions were calculated where there were radar observations from all three of these antennas, whereas ERS scatterometers were sometimes forced to use observations from only two antennas (Zeccheto *et al.* [19]). For fixed-antenna scatterometers (SeaSat, ERS-1/2, and NSCAT), the use of three or more antennas is essential for accurate determination of the wind direction (Naderi *et al.* [17]). The beam arrangement on SeaWinds instruments is a new design, with two conically rotating beams at fixed incidence angles. This design allows a single very wide observational swath. This scanning geometry has four substantially different angles over the portion of the swath similar to the NSCAT's coverage. However, near nadir and near the edges of the swath the angles are similar, resulting in decreased accuracy in these parts of the swath. Furthermore, only one of the two beams reaches the outer 100 km of the swath. These problems are somewhat compensated by a much greater observation density. NSCAT had three or four observations within a 25 × 25 km cell, whereas SeaWinds typically has between 8 and 15 observations within its 25 × 25 km cells.

The functions describing the wind direction are sinusoidal. Combining these functions to minimize the misfit usually results in multiple minima (ambiguous solutions often called "ambiguities"). Ideally, for fixed-antenna scatterometers, the best fit corresponds to the correct direction, the next-best fit is in approximately the opposite direction, and the next two minima are in directions roughly perpendicular to the wind direction. For SeaWinds scatterometers, the solution geometry varies across the swath. The solutions are similar to fixed-antenna scatterometer solutions in the part of the swath similar to NSCAT coverage, but differ greatly near nadir and near the swath edges. The process of choosing a direction from the multiple possible solutions is called ambiguity selection. Noise and spatial/temporal variability can change the quality of fit and thereby cause incorrect directions (also known as aliases) to be chosen. NSCAT's ambiguity removal skill was further improved by using two polarizations with one antenna, whereas SeaWinds ambiguity selection is improved by greater observation density. For NSCAT and QuikSCAT winds, a median filter (applied to ambiguity selection rather than wind direction) is also used to improve ambiguity selection.

An additional challenge is rain. Rain influences radar returns through three processes: backscatter from rain drops, attenuation of the signal passing through the rain (Weissman *et al.* [20]), and modification of the sea surface shape by raindrop impacts (Bliven *et al.* [21]; Sobieski and Bliven [22]; Sobieski *et al.* [23]). The influence of these processes on the accuracy of winds is a function of scatterometer design. Rain has a greater influence at large incidence angles (the signal passes through more rain) and for Ku-band radar (NSCAT and SeaWinds) rather than C-band (ERS-1/2). Rain is not considered a serious problem for the ERS scatterometers. For NSCAT, rain contributed to substantial errors in the outer parts of the swaths, whereas for SeaWinds observations rain can have a substantial influence throughout the swath. Modeling these problems is a concern in ongoing research (Weissmann *et al.* [20, 24]; Yueh *et al.* [25]; Draper and Long [26]).

## 2.2 Calibration and equivalent neutral winds

Planning of post-SeaSat scatterometer missions assumed that scatterometers respond directly to surface turbulent stress rather than wind speed (O'Brien *et al.* [2]). The lack of sufficient calibration points for the direct determination of stress was foreseen, and it was decided to calibrate the instrument for the observation of surface (10 m) wind. Colocated scatterometer and buoy winds were used to calibrate the scatterometer (Freilich and Dunbar [27]; Wentz and Smith [28]). More elaborate techniques were developed to ensure that a common reference scale was used to measure winds from three colocated types of observations (Stoffelen [29]). Winds determined with the NSCAT-1 geophysical model function (Wentz and Smith [28]) have been validated against a wide range of *in situ* and remotely sensed winds, and SeaWinds data have been validated with observations from research vessels (Bourassa *et al.* [30]).

Scatterometer wind speeds have been calibrated to 10 m "equivalent neutral wind speeds" (Cardone [31]; Ross *et al.* [32]; Cardone *et al.* [33]; Verchell *et al.* [34]), which differ from surface wind speeds in a manner believed to be consistent with the physics to which the scatterometer responds. The differences can easily be explained with the equation for the modified log-wind profile:

$$\boldsymbol{U}(z) - \boldsymbol{U}_{\mathrm{sfc}} = (\boldsymbol{u}_*/k)\,[\,\ln(z/z_0) - \varphi(z, z_0, L)], \tag{1}$$

where $\boldsymbol{U}$ is the vector wind, $\boldsymbol{U}_{\mathrm{sfc}}$ is the velocity frame of reference (the surface current), $\boldsymbol{u}_*$ is the friction velocity, $k$ is von Karman's constant, $z$ is the height above the local mean surface (10 m in this case), $z_0$ is the roughness length, $\varphi$ is a function of atmospheric stability, and $L$ is the Monin–Obukhov scale length which is a measure of atmospheric stability. The response of scatterometers to the sea surface ($z \approx 0$), and the stability term ($\varphi$) is largely a function of $z/L$. Therefore, the accepted approach is to eliminate the stability term in the height adjustment. The equivalent neutral wind speed is parameterized in a manner similar to eqn (1), and uses the same nonneutral values of $\boldsymbol{u}_*$ and $z_0$; however, the stability term ($\varphi$) is set to zero, implying

$$\boldsymbol{U}_{\mathrm{EN}}(z) - \boldsymbol{U}_{\mathrm{sfc}} = (\boldsymbol{u}_*/k)\ln(z/z_0). \tag{2}$$

Hence, the differences between $\boldsymbol{U}_{10\mathrm{EN}}$ and $\boldsymbol{U}_{10}$ are stability dependent,

$$\boldsymbol{U}_{10\mathrm{EN}} - \boldsymbol{U}_{10} = \boldsymbol{u}_*\varphi(10, z_0, L)/k. \tag{3}$$

The kinematic surface turbulent stress is equal to the square of the friction velocity, and the friction velocity used in calibration of scatterometer winds should be the nonneutral value rather than the neutral value. However, values of $|\boldsymbol{U}_{10\text{EN}} - \boldsymbol{U}_{10}|$ are usually much less than 0.5 m s$^{-1}$, and relatively large values of $|\boldsymbol{U}_{10\text{EN}} - \boldsymbol{U}_{10}|$ tend to be associated with very stable stratification.

### 2.3 Determining stress from scatterometer winds

The calculation of surface turbulent stress ($\boldsymbol{\tau}$) from wind speed also requires knowledge of the atmospheric stratification [eqn (1)]

$$\boldsymbol{\tau} = \rho \boldsymbol{u}_* |\boldsymbol{u}_*|. \tag{4}$$

The influence of atmospheric stratification on surface stress can be very large, particularly for stable stratification. The calculation of atmospheric stratification parameters requires knowledge of the air/sea temperature difference. Satellites can accurately determine sea surface skin temperature; however, for subsynoptic time scales they are not yet effective in determining the temperature 10 m above the surface.

One of the great advantages of equivalent neutral winds is that no knowledge of the atmospheric stability is required to determine surface stress. All that is needed is the density of the air and a neutral drag coefficient.

$$\boldsymbol{\tau} = \rho C_D \boldsymbol{U}_{10\text{EN}} |\boldsymbol{U}_{10\text{EN}}|. \tag{5}$$

The difficulties with this approach are the determination of the drag coefficient (described in Section 3) and calibration of $\boldsymbol{U}_{10\text{EN}}$ for severe weather. The calibration becomes an issue for wind speeds >20 m s$^{-1}$ due to the lack of calibration data, and because of the interpretation of scatterometer winds. For wind speeds >16 m s$^{-1}$, wave-related surface motion could become important in interpreting scatterometer winds (Section 6).

## 3 *In situ* data

A preliminary analysis of observations from the Storm Wave Study experiment (SWS-2) was kindly provided by Peter K. Taylor (see also Dobson *et al.* [35]; Taylor *et al.* [36]). These ship-based observations were gathered in the North Atlantic Ocean, with the goal of gathering a high quality data set for severe wind conditions. A key advantage of the data set is the existence of colocated buoy wave data. These observations have the additional advantage of covering a large range of wind conditions: 4–24 m s$^{-1}$, with the quality control criteria described below. It is also the data set that Taylor and Yelland [14] used to demonstrate that models using the HEXOS parameterization (described by Smith *et al.* [10]) overestimated stresses for high wind speeds.

The following constraints were applied to the SWS-2 observations:

1. the sonic anemometer's estimate of the speed of sound is between 332 and 338 m s$^{-1}$,
2. the standard deviation of the platform's heading is less than 14°,
3. the dimensionless Monin–Obukhov scale length [$z/L$; eqns (4–6)] is less than 0.18.

These criteria serve as quality control on the stress and wave observations, as well as removing conditions that were too far from neutral to be considered. The first constraint ensures that the speed of sound, which is critical to the calculation of turbulent fluxes, is within the range expected from the observed air temperatures (Taylor [37]). The second constraint attempts to remove cases where excessive ship motion or flow distortion introduces too much variability in the observations: approximately 1% of the observations are removed due to this constraint (Taylor [37]). The third constraint applies a measure of atmospheric stratification ($z/L$) to remove cases that are far from neutral stability, where the parameterizations that account for such departures are relatively uncertain.

## 4 Flux model

The fluxes considered in this model are the downward momentum flux ($\boldsymbol{\tau}$), and the upward surface turbulent fluxes of sensible ($H$), moisture ($E$), and latent heat ($Q$). Stress can be modeled in terms of the friction velocity ($\boldsymbol{u}_*$):

$$\boldsymbol{\tau} = \rho \boldsymbol{u}_* |\boldsymbol{u}_*|, \tag{6}$$

where $\rho$ is the density of the air. Sensible heat, moisture, and latent heat fluxes are

$$H = -\rho C_p \theta_* |\boldsymbol{u}_*|, \tag{7}$$

$$E = -\rho q_* |\boldsymbol{u}_*|, \tag{8}$$

$$Q = -\rho L_{\mathrm{v}} q_* |\boldsymbol{u}_*| = L_{\mathrm{v}} E, \tag{9}$$

where $\theta_*$ and $q_*$ are scaling parameters analogous to $u_*$, $C_p$ is the specific heat of air, and $L_{\mathrm{v}}$ is the latent heat of vaporization. The following considerations improve the accuracy of these modeled surface turbulent fluxes by improving the accuracy of the modeled values of $u_*$.

The direct influence of surface waves on flux and airflow characteristics ($\boldsymbol{u}_*$ and $z_0$) is determined by the relation between $\boldsymbol{u}_*$ and roughness length ($z_0$). Given $z_0(\boldsymbol{u}_*)$ and the modified log-wind relation $\boldsymbol{U}(z)$, where $z$ is the height above the local mean surface, it is possible to iteratively solve for $\boldsymbol{u}_*(\boldsymbol{U})$ and $\boldsymbol{\tau}(\boldsymbol{U})$. The modified log-wind relation is

$$\boldsymbol{U}(z) - \boldsymbol{U}_{\mathrm{s}} = \frac{\boldsymbol{u}_*}{k} \ln \frac{z-d}{z_0} + 1 + (z, z_0, L), \tag{10}$$

where $k$ is von Kârmân's constant, $d$ is the displacement height (the height at which the log-wind profile extrapolates to zero wind speed), and $L$ is the Monin–Obukhov stability length. The influence of atmospheric stratification in the boundary layer is modeled through the Monin–Obukhov stability length (Liu *et al.* [38]). The profiles of potential temperature ($\theta$) and specific humidity ($q$) have functional forms similar to the log-wind profile. The parameterization of momentum roughness length [eqn (12)] is adapted from Bourassa [13], and the roughness lengths for potential temperature ($z_{0\theta}$) and specific humidity ($z_{0q}$) are adapted from the surface renewal model of Clayson *et al.* [39], which implies that they are proportional to the momentum roughness length. The parameterization of $L$ is identical to that used in the BVW (Bourassa–Vincent–Wood) flux model [12], the CFC (Clayson–Fairall–Curry) model [39], and also Bourassa [13].

## 4.1 Momentum roughness length

There have been many attempts to model the dependence of stress on sea state. Many of these studies have followed the suggestions of Charnock [40], who used dimensional analysis to parameterize momentum roughness length:

$$z_0 = a\,u_*^2/g, \tag{11}$$

where $a$ is a dimensionless parameter (now called Charnock's constant) that is often considered to be a function of sea state conditions. The wave dependence in subsequent models of surface turbulent stress typically enters through Charnock's constant in the gravity wave contribution to the roughness length parameterization (see Kusaba and Masuda [5]; Geernaert [6]; Toba *et al.* [7]; Perrie and Toulany [8]; Maat *et al.* [9]; Smith *et al.* [10]; Bourassa *et al.* [12]). The momentum roughness length parameterization used here (Bourassa [13]) is a modification of that of BVW [12]. BVW stresses were found to greatly overestimate the high wind speed stresses observed in SWS-2 as discussed by Dobson *et al.* [35], Talyor *et al.* [36], and Taylor [41]. It was speculated (Bourassa *et al.* [12]; Taylor [41]) that these biases were due to extrapolation of the HEXOS parameterization for momentum roughness length (Smith *et al.* [10]) to conditions for which it was not valid. The roughness length parameterization used here, given in eqn (12), can be written with no explicit dependence on sea state, where the gravity wave roughness length is a two-dimensional version ($i = 1, 2$) of Charnock's relation [40]

$$z_{0_i} = \left[\beta_v' \frac{0.11\nu}{|u_{*_i}|} + \left[\left(\beta_c' \frac{b\sigma}{\rho_{\mathrm{w}}|\boldsymbol{u}_*||\boldsymbol{u}_* \cdot \hat{\boldsymbol{e}}_i|}\right)^2 + \left(\beta_g' \frac{a|\boldsymbol{u}_*||\boldsymbol{u}_* \cdot \hat{\boldsymbol{e}}_i|}{g}\right)^2\right]^{0.5}\right]. \tag{12}$$

Here, the $\beta$ terms are binary weights for the roughness lengths (from left to right) associated with contributions to surface roughness from three types of surface features: an aerodynamically smooth surface (Nikuradse [42]; Kondo [43]), capillary waves (Bourassa *et al.* [12]), and gravity waves (Smith *et al.* [10]), where $\nu$ is the molecular viscosity of air, $b$ is a dimensionless constant (determined from laboratory observations), $\sigma$ is the surface tension, $\rho_{\mathrm{w}}$ is the density of water, and $g$ is the gravitational acceleration. The influence of sea state enters solely through the modification of vertical shear in wind speed, given in eqn (13), due to a nonzero lower boundary condition, the wave-induced surface motion

$$[U(z) - fU_{\text{orb}} - U_{\text{current}}]\,\hat{e}_i = \frac{u_* \hat{e}_i}{k} \ln \frac{z-d}{z_{0_i}} + 1 + (z, z_0, L). \qquad (13)$$

The roughness length is therefore anisotropic, with unit vectors parallel ($\hat{e}_1$) and perpendicular ($\hat{e}_2$) to the mean direction of wave motion. The orbital velocity ($U_{\text{orb}}$) term transforms the velocity frame of reference to that of a fraction $f$ of the orbital velocity of the dominant waves. Laboratory studies (Okuda *et al.* [44]) have shown that most of the interactions between wind and wind-driven waves occur near the crest of the dominant waves. This process is modeled by modifying the lower boundary condition for velocity. The orbital speed of gravity waves is approximated by

$$U_{\text{orb}} = \pi H_s / T_p, \qquad (14)$$

where $H_s$ is the significant wave height and $T_p$ is the corresponding significant wave height period. For gravity waves, $U_{\text{orb}}$ can easily be related to the significant slope used by Taylor and Yelland [14]. The fraction of the orbital velocity ($f$) that modifies the surface wind was set at 80%, following Bourassa [13]. The value of 0.8 applies to wind-driven waves: flow separation can be much less important for swell than for wind-driven waves, suggesting that the value for $f$ should be smaller for most swell conditions.

The use of significant wave height in eqn (14) is an oversimplification: $H_s$ can be a combination of wind waves and swell from multiple sources. Directional differences in the wave spectra will contribute to errors in the estimated orbital velocity and, consequently, to errors in the stresses. However, wave spectra are rarely available from observational studies, whereas $H_s$ is frequently recorded in *in situ* observations and is remotely sensed through satellite altimetry. In practice, there are few sources of spectral wave information other than from wave models (e.g., WAM model outputs as noted by Günther *et al.* [45], and Janssen [46, 47]) as implemented in operational weather centers. The lack of wave directional information in the SWS-2 data, and hence the lack of consideration in this study, presumably causes a substantial fraction of the unaccounted variability in the SWS-2 data.

Bourassa [13] pointed out that the roughness length could be given a more conventional form by moving the $fU_{\text{orb}}$ term from the left hand side of eqn (13) to the roughness length in eqn (12). However, this is inappropriate here because the roughness lengths for potential temperature and moisture are dependent on the roughness length for momentum. Comparisons of modeled and observed sensible and latent heat fluxes (not shown) support the form of roughness length used in eqn (13).

### 4.2 Displacement height

The displacement height is not considered as another free parameter: it is derived as part of the lower boundary condition on the log-wind profile. Several derivations of the modified log-wind profile lead to eqn (15) after integrating from the smallest length scale of eddies to the length scale associated with an arbitrary height $z$

$$\frac{\partial \bar{u}}{\partial z} = \frac{\sqrt{\tau/\rho}}{k_{\mathrm{v}}} \frac{1}{(z - z_0)}. \tag{15}$$

The modified log-wind profile is then found by integrating eqn (15) from the height where the surface relative wind speed extrapolates to zero to the wind speed at a height $z$

$$\bar{u}\Big|_{z=d}^{z=z} = \frac{\sqrt{\tau/\rho}}{k_{\mathrm{v}}} \log\left(\frac{z}{z_0} + 1\right)\Bigg|_{z=d}^{z=z}. \tag{16}$$

Hence the bottom boundary condition $\bar{u}(d)$ corresponds to the frame of reference for the wind speed. This frame of reference was found to be approximately 80% of the magnitude of the orbital velocity of the dominant wind waves (Bourassa [13]). The displacement height is then the height that corresponds to this speed: 80% of the significant wave height ($fH_{\mathrm{s}}$) for wind-driven waves. This value of the displacement height follows directly from the derivation of the modified log-wind profile and knowledge of the frame of reference for wind speed. The orbital velocity term usually reduces the stress by decreasing the wind shear. In contrast, the displacement height acts to increase the wind shear.

## 5 Comparisons with observations

The model is evaluated with SWS-2 observations, because this data set is of the highest quality and should therefore provide an important calibration. In particular, all the required meteorological data, flux data, and wave data were recorded (although directional wave information is not available for this study). As the surface water is well mixed, the differences between bulk and skin temperature can be safely ignored. These are important considerations because they remove substantial biases in both friction velocity and the sea surface temperature.

Charnock's constant is highly dependent on the velocity frame of reference ($f\boldsymbol{U}_{\mathrm{orb}} + \boldsymbol{U}_{\mathrm{current}}$) and the displacement height ($d$): a small percentage change in $|\boldsymbol{u}_*|$ corresponds to a large percentage change in $z_0$. This approach reduces the root mean square (rms) differences between modeled and observed friction velocity from 0.078 to 0.041 m s$^{-1}$. Displacement height was not considered by Bourassa [13], and it was found that $f = 0.8$ and $a = 0.064$ resulted in a good fit to the SWS-2 data, particularly in the mean. However, this result was highly influenced by observations at extreme wind speeds ($U_{10} > 15\,\mathrm{m\,s^{-1}}$), where the displacement height was relatively significant. A different value for Charnock's constant would be found if the extremely strong wind speeds or large wave heights were excluded. The change in modeled values of $u_*$ due to this uncertainty in the value of Charnock's constant is small for SWS-2 data with $7 < U_{10} \leq 15\,\mathrm{m\,s^{-1}}$. Although the application of any model to conditions for which it was not tested

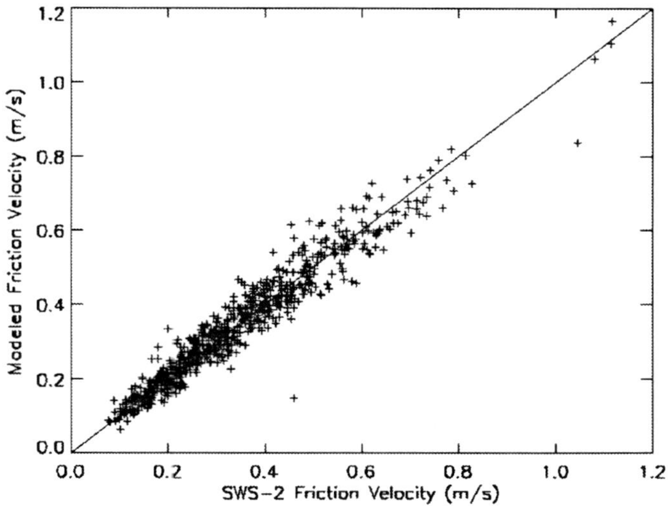


Figure 1: Comparison of modeled and observed SWS-2 friction velocity magnitudes shown as a scatter plot.

is always problematic, it is necessary in the absence of appropriate data. Models with more detailed physical considerations are more likely to result in an accurate extrapolation into untested conditions. Here, consideration of $d$ results in $a = 0.035$ (Fig. 1), with a very good match to the means shown in Fig. 2 except for the most extreme conditions where the sample size is relatively small. Furthermore, these values for $f$ and $a$ are found for any reasonably large subrange of wind speeds or friction velocities from the SWS-2 data.

There is considerable spread about the means (Fig. 2), which is expected in part due to the lack of directional sea state information. The rms difference between modeled and observed values is 0.041 m s$^{-1}$, when the orbital velocity and displacement height are considered, and increases to 0.078 m s$^{-1}$ when these considerations are ignored. The median value of the absolute value of the relative error is 11% when orbital velocity and displacement height are considered. Remarkably, the consideration of orbital velocity has also been demonstrated by Bourassa *et al.* [12] to account for observed variability in surface stress for conditions of low wind speeds and old swell, as noted by Donelan *et al.* [48].

The influences of $u(d)$ and $d$ on the friction velocity are demonstrated with the SWS-2 data in Fig. 3. Although the wind speed modification tends to dominate for wind speeds throughout the range of conditions shown, the influence of this modification in the SWS-2 data is approximately a constant proportion of the magnitude of the friction velocity. The influence of displacement height is very small (Fig. 3) for friction velocities below approximately 0.55 m s$^{-1}$ (wind speeds below approximately 12 m s$^{-1}$). However, it increases with increasing SWS-2 friction velocity (and increasing wave heights).

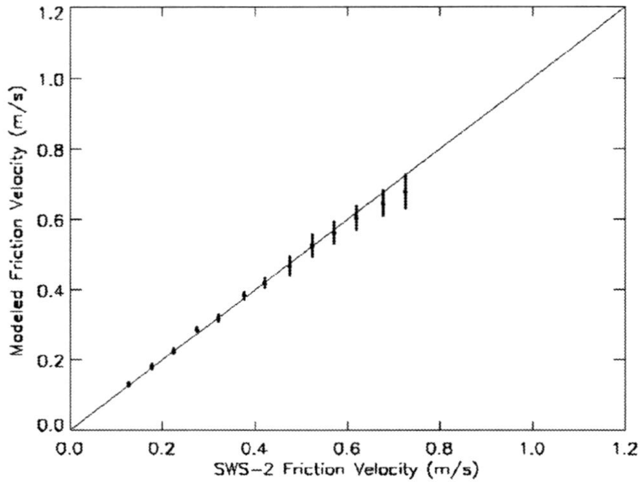


Figure 2: Comparison of modeled and observed SWS-2 friction velocity magnitudes, shown as means and uncertainty in those means. The error bars indicate ±3 standard deviations in the mean. There are only 10 points in each of the two bins for the extreme means, whereas there are 22–104 points in the other bins. The bin width for SWS-2 $u_*$ values is 0.05 m s$^{-1}$, and means are calculated only when there are at least eight values in a bin.

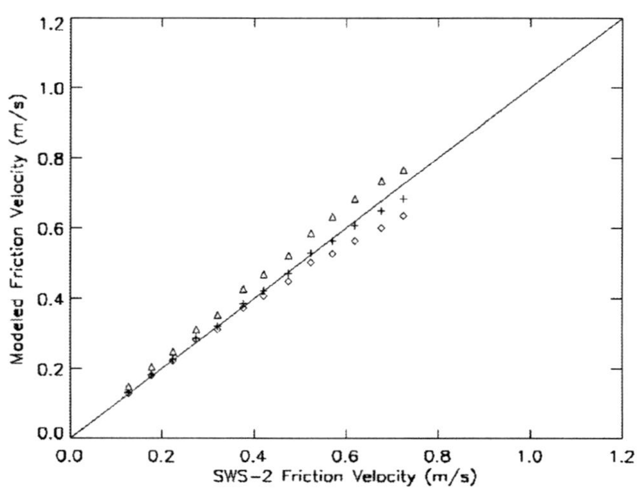


Figure 3: Comparison of modeled influences of orbital velocity and displacement height on friction velocity in the SWS-2 data. The modeled friction velocity considering both influences (denoted +) is approximately 10 to 12% less than when orbital velocity is not considered (denoted △); and is slightly larger than when displacement height is ignored (denoted ◇).

## 6 Applications to scatterometry

One of the early goals of scatterometry was to remotely sense ocean vector surface stress. The above results indicate that this goal might be close to being realized. The calibration of radar backscatter to equivalent neutral winds deals with the dependence of stress on atmospheric stability. However, wave characteristics also influence surface turbulent stress. The dependence of the calibration on the horizontal vector of the dominant waves' orbital velocity accounts for much of the dependency on wave characteristics. Scatterometers clearly respond to wind speed minus a fraction of the orbital velocity; however, it is not certain that this fraction is equal to $f$ [from eqn (10)], which describes the influence of waves on the relation between earth-relative winds and stress. If the influence of orbital velocity on scatterometer wind speeds is similar in magnitude to $f$, then scatterometers can easily be used to estimate surface stress as well as the surface stress derivatives that are of great importance to ocean forcing. If the above case occurs, then this estimate of stress will be accurate so long as the influence of displacement height can be ignored. The overestimation that occurs due to ignoring the displacement height is less than 10% for SWS-2 wind speeds less than approximately $20\,\mathrm{m\,s^{-1}}$. This range of wind speeds accounts for the vast majority of ocean winds, including most winds associated with storms.

It seems likely that the mean influence of displacement height could be considered to further reduce bias in satellite-based estimates of surface stress for extreme conditions. Ideally, surface stress and wave observations could be obtained for such conditions so that the above model could be either validated or improved for extreme conditions. It is interesting to note that the orbital velocity modification could account for several seemingly conflicting observations in hurricanes: surface winds that are extrapolated downward are found to have greater values than are consistent with a log-wind profile with zero surface wind and the associated estimated surface stress. The orbital velocity could account for a much larger than expected near-surface wind speed, for such extreme conditions.

## 7 Conclusions

The influence of wave characteristics on surface stress has been examined. Previous approaches have modified the momentum roughness length; however, the approach described here allows for a simple Charnock parameterization of momentum roughness length. The influences of waves are considered in the lower boundary condition in the modified log-wind profile. The horizontal component of the dominant waves' orbital velocity modifies the near-surface boundary condition for wind velocity. Eighty percent of this horizontal component is vector-subtracted from the near-surface wind speed. Typically, this modification decreases the vertical wind shear and thereby also decreases the wind stress. However, if the waves are propagating against the wind direction there will be an increase in the wind shear and an increase in the surface turbulent stress. This boundary condition on velocity is matched with a boundary condition related to the vertical coordinate system, implying a nonzero displacement height. This knowledge provides the information for a first estimate of the displacement height, representing a vertical offset of the log-wind profile. The displacement height for wind waves is found to be 80% of the significant wave height. Some caution must be taken in applying these results to conditions where the waves are dominated by swell. The influence of the orbital velocity on stress has been verified with observations; however, the value of the displacement height has not been adequately studied for these conditions. Nevertheless, for very high wind speeds, the consideration of displacement height helps to reduce a systematic underestimate of the surface turbulent stress.

## Acknowledgments

I thank Peter Taylor for providing the SWS-2 data and advice on quality assurance. COAPS receives its base funding from ONR's Secretary of Navy Grant to James J. O'Brien. Current support is from NASA's Ocean Vector Winds Science Team, OSU's SeaWinds project, Florida State University's First Year Investigator Program, NOPP, and the NSF.

## References

[1] Schlax, M.G., Chelton, D.B. & Freilich, M.H., Sampling errors in wind fields constructed from single and tandem scatterometer datasets. *Journal of Atmospheric and Oceanic Technology*, **18**, pp. 1014–1036, 2001.

[2] O'Brien, J.J. *et al.*, *Scientific opportunities using satellite wind stress measurements over the ocean.* Report of the S3 Working Group, Nova University Press, 1982.

[3] Powell, M.D., Vickery, P.J. & Reinhold, T.A., Reduced drag coefficient for high wind speeds in tropical cyclones. *Nature*, **422**, pp. 279–283, 2003.

[4] Chen, S. Personal communication, 2003.

[5] Kusaba, T. & Masuda, A., The roughness height and drag law over the water surface based on the hypothesis of local equilibrium. *Journal of the Oceanographical Society of Japan*, **44**, pp. 200–214, 1988.

[6] Geernaert, G.L., Measurements of the angle between the wind stress vector in the surface layer over the North Sea. *Journal of Geophysical Research*, **91**, pp. 7667–7679, 1990.

[7] Toba, Y., Ida, N., Kawamura, H., Ebuchi, N. & Jones, I.S.F., Wave dependence of sea-surface wind stress. *Journal of Physical Oceanography*, **20**, pp. 705–721, 1990.

[8] Perrie, W. & Toulany, B., Fetch relations for wind-generated waves as a function of wind-stress scaling. *Journal of Physical Oceanography*, **20**, pp. 1666–1681, 1990.

[9] Maat, M., Kraan, C. & Oost, W.A., The roughness of wind waves. *Boundary-Layer Meteorology*, **54**, pp. 89–103, 1991.

[10] Smith, S.D., Anderson, R.J., Oost, W.A., Kraan, C., Maat, N., DeCosmo, J., Katsaros, K.B., Davidson, K.L., Bumke, K., Hasse, L. & Cadwick, H.M., Sea surface wind stress and drag coefficients: the HEXOS results. *Boundary-Layer Meteorology*, **60**, pp. 109–142, 1992.

[11] Yelland, M.J., Moat, B.I., Taylor, P.K., Pascal, R.W., Hutchings, J. & Cornell, V.C., Wind stress measurements from the open ocean corrected for airflow distortion by the ship. *Journal of Physical Oceanography*, **28**, pp. 1511–1526, 1998.

[12] Bourassa, M.A., Vincent, D.G. & Wood, W.L., A flux parameterization including the effects of capillary waves and sea state. *Journal of the Atmospheric Sciences*, **56**, pp. 1123–1139, 1999.

[13] Bourassa, M.A., An improved sea state dependency for surface stress derived from *in situ* and remotely sensed winds. *Advances in Space Research*, **33(7)**, pp. 1136–1142, 2004.

[14] Taylor, P.K. & Yelland, M.J., The dependence of sea surface roughness on the height and steepness of the waves. *Journal of Physical Oceanography*, **18**, pp. 572–590, 2001.

[15] Quilfen Y., Chapron, B. & Vandemark, D., The ERS scatterometers wind measurement accuracy: evidence of seasonal and regional biases. *Journal of Atmospheric and Oceanic Technology*, **18**, pp. 1684–1697, 2001.

[16] Katsaros, K.B. & Brown, R.A., Legacy of SeaSat mission for studies of the atmosphere and air-sea-ice interactions. *Bulletin of the American Meteorological Society*, **72**, pp. 967–981, 1991.

[17] Naderi, F.M., Freilich, M.H. & Long, D.G., Spaceborne radar measurements of wind velocity over the ocean—an overview of the NSCAT scatterometer system. *Proceedings of the IEEE*, **79**, pp. 850–866, 1991.

[18] Freilich, M.H. & Long, D.G. Personal communication, 1998.

[19] Zeccheto, S., Tanduo, I. & Quilfen, Y., The CERSAT dealiasing of ERS-2 scatterometer winds in the Mediterranean Sea. *CERSAT News*, Scientific topic number 2, pp. 1–4, 1999.

[20] Weissman, D.E., Bourassa, M.A. & Tongue, J., Effects of rain-rate and wind magnitude on SeaWinds scatterometer wind speed errors. *Journal of Atmospheric and Oceanic Technology*, **19**, pp. 738–746, 2002.

[21] Bliven, L.F., Branger, H., Sobieski, P.C. & Giovanangeli, J.-P., An analysis of scatterometer returns from a water surface agitated by artificial rain. *International Journal of Remote Sensing*, **14**, pp. 2315–2329, 1993.

[22] Sobieski, P. & Bliven, L.F., Analysis of high speed images of raindrop splash products and ku-band scatterometer returns. *International Journal of Remote Sensing*, **16**, pp. 2721–2726, 1995.

[23] Sobieski, P.W., Craeye, C. & Bliven, L.F., Scatterometric signatures of multivariate drop impacts on fresh and salt water surfaces. *International Journal of Remote Sensing*, **20**, pp. 2149–2166, 1999.

[24] Weissman, D.E., Bourassa, M.A., Tongue, J. & O'Brien, J.J., Calibrating the QuikSCAT/SeaWinds radar for measuring rain over water. *IEEE Transactions on Geoscience and Remote Sensing*, **41**, pp. 2814–2820, 2003.

[25] Yueh, S.H., Styles, B. & Liu, W.T., QuikSCAT wind retrievals for tropical cyclones. *IEEE Transactions on Geoscience and Remote Sensing*, **41**, pp. 2616–2628, 2003.

[26] Draper, D.W. & Long, D.G., Evaluating the effect of rain on SeaWinds scatterometer measurements. *Journal of Geophysical Research*, **109**(C02005), doi:10.1029/2002JC001741, 2004.

[27] Freilich, M.H. & Dunbar, R.S., The accuracy of the NSCAT 1 vector winds: comparison with National Data Buoy Center buoys. *Journal of Geophysical Research*, **104**, pp. 11231–11246, 1999.

[28] Wentz, F.J. & Smith, D.K., A model function for the ocean-normalized radar cross section at 14 GHz derived from NSCAT observations. *Journal of Geophysical Research*, **104**, pp. 11499–11514, 1999.

[29] Stoffelen, A., Toward the true near-surface wind speed: error modeling and calibration using triple collocation. *Journal of Geophysical Research*, **103**, pp. 7755–7766, 1998.

[30] Bourassa, M.A., Legler, D.M., O'Brien, J.J. & Smith, S.R., SeaWinds validation with research vessels. *Journal of Geophysical Research*, **108**, 3019, doi:10.1029/2001JC001081, 2003.

[31] Cardone, V.J., Specification of the wind distribution in the marine boundary layer for wave forecasting. *Geophys. Sci. Lab. N.Y. Univ. Report TR-69-1* (NTIS AD 702–490), 1969.

[32] Ross, D.B., Cardone, V.J., Overland, J., McPherson, R.D., Pierson, W.J., Jr. & Yu, T., Oceanic surface winds. *Advances in Geophysics*, **27**, pp. 101–138, 1985.

[33] Cardone, V.J., Jensen, R.E., Resio, D.T., Swail V.R. & Cox, A.T., Evaluation of contemporary ocean wave models in rare extreme events: the "Halloween Storm" of October 1991 and the "Storm of the Century" of March 1993. *Journal of Atmospheric and Oceanic Technology*, **13**, pp. 198–230, 1996.

[34] Verschell, M.A., Bourassa, M.A., Weissman, D.E. & O'Brien, J.J., Model validation of the NASA scatterometer winds. *Journal of Geophysical Research*, **104**, pp. 11359–11374, 1999.

[35] Dobson, F.W., Anderson, R.J., Taylor, P.K. & Yelland, M.J., Storm Wind Study II: open ocean wind and sea state measurements. *Proc. Symp. on the Wind-Driven Air-Sea Interface: Electromagnetic and Acoustic Sensing, Wave Dynamics and Turbulent Fluxes*, ed. M.L. Banner, University of New South Wales, pp. 295–296, 1999.

[36] Taylor, P.K., Yelland, M.J., Dobson, F.W. & Anderson, R.J., Storm Wind Study II: wind stress estimates from buoy and ship. *Proc. Symp. on the Wind-Driven Air-Sea Interface: Electromagnetic and Acoustic Sensing, Wave Dynamics and Turbulent Fluxes*, ed. M.L. Banner, University of New South Wales, pp. 353–354, 1999.

[37] Taylor, P.K. Personal communication, 2003.

[38] Liu, W.T., Katsaros, K.B. & Businger, J.A., Bulk parameterization of air-sea exchanges of heat and water vapor including the molecular constraints at the interface. *Journal of the Atmospheric Sciences*, **36**, pp. 1722–1735, 1979.

[39] Clayson, C.A., Fairall, C.W. & Curry, J.A., Evaluation of turbulent fluxes at the ocean surface using surface renewal theory. *Journal of Geophysical Research*, **101**, pp. 28503–28513, 1996.

[40] Charnock, H., Wind stress on a water surface. *Quarterly Journal of the Royal Meteorological Society*, **81**, pp. 639–640, 1955.

[41] Taylor, P.K. Personal communication, 1999.

[42] Nikuradse, J., *Stromungsgesetze in rauben Rohren*, V.D.I. Forschungsheft, p. 361, 1933.

[43] Kondo, J., Air–sea bulk transfer coefficients in diabatic conditions. *Boundary-Layer Meteorology*, **9**, pp. 91–112, 1975.

[44] Okuda, K., Kawai, S. & Toba, Y., Measurements of skin friction distribution along the surface of wind waves. *Journal of the Oceanographical Society of Japan*, **33**, pp. 190–198, 1977.

[45] Günther, H. *et al.*, *Technical Report No. 68: Implementation of a third generation ocean wave model at the European Centre for Medium-Range Weather Forecasts*, European Centre for Medium-Range Weather Forecasts, p. 34, 1992.

[46] Janssen, P.A.E.M., Wave-induced stress and the drag of airflow over sea waves. *Journal of Physical Oceanography*, **19**, pp. 745–754, 1989.

[47] Janssen, P.A.E.M., Quasi-linear theory of wind-wave generation applied to wave forecasting. *Journal of Physical Oceanography*, **21**, pp. 1631–1642, 1991.

[48] Donelan, M.A., Drennan, W.M. & Katsaros, K.B., The air-sea momentum flux in conditions of wind sea and swell. *Journal of Physical Oceanography*, **27**, pp. 2087–2099, 1997.

# CHAPTER 3

# Similarity and dissimilarity in air–sea momentum and $CO_2$ transfers: the nondimensional transfer coefficients in light of the windsea Reynolds number

Y. Toba[1,2,3], S. Komori[4], Y. Suzuki[5] & D. Zhao[6]
[1] *Tohoku University (Emeritus), Japan.*
[2] *Earth Observation Research Center, National Space Development Agency of Japan (EORC/NASDA), Japan.*
[3] *Research Institute for Environmental Sciences and Public Health of Iwate Prefecture (I-RIEP), Japan.*
[4] *Department of Mechanical Engineering and Advanced Research Institute of Fluid Science and Engineering, Kyoto University, Japan.*
[5] *Research Department, Japan Weather Association, Japan.*
[6] *Physical Oceanography Laboratory, Ocean University of China, China.*

## Abstract

From dimensional considerations it is deduced that in *wind–windsea equilibrium* the fluid dynamical conditions can be described by only two nondimensional parameters. One is the wave age for the representative windsea. This describes the degree of windsea development relative to the local wind. The other is the *windsea Reynolds number* $R_B$, which is defined by $u_*^2/\nu\sigma_p$, where $u_*$ is the air friction velocity, $\nu$ is the air kinematic viscosity, and $\sigma_p$ is the angular frequency at the windsea spectral peak. $R_B$ was introduced by Toba and Koga in 1986, and formerly it was denoted as the *breaking-wave parameter*. It is the fundamental controlling parameter for the *windsea boundary layer*, and consequently for the behavior of air–sea transfers. $R_B$ is shown to be much superior to the 10-m wind speed, $U_{10}$ or $u_*$, in studying the behavior of transfer coefficients.

The air–sea $CO_2$ transfer has traditionally been expressed in terms of a transfer velocity $k_L$ as a function of $U_{10}$. We propose the use of a nondimensional transfer velocity $k_L/U_{10}$, motivated by the drag coefficient $C_D$ for the momentum transfer. Existing data show that there is fundamental similarity in the behavior of these nondimensional coefficients, with some dissimilarity in a range of $R_B$. Detailed findings are as follows.

1. Though $C_D$ generally increases with $R_B$, laboratory data show two troughs, or critical points at $R_B = 2 \times 10^2$ and $10^3$, with a flat bulge region in between. The first trough corresponds to the transition from the region where tangential stress prevails to the occurrence of airflow separation and reattachment to create the form drag. $R_B = 10^3$ is a conspicuous critical point beyond which windsea breaking occurs on the fully rough water surface, namely it is the transition to the full turbulence regime of the windsea boundary layer.
2. Between $R_B = 2 \times 10^2$ and $10^3$, $k_L/U_{10}$ decreases. The reason is that in gas transfer processes, the mechanism corresponding to the pressure–force for the momentum transfer is not available.
3. Field data, under the usual swell conditions, show smaller values or a remarkable shift of the second critical point to larger values in $R_B$, for both the nondimensional transfer coefficients. This corresponds to the fact that windsea energy becomes small in the presence of swell in some conditions, causing the sea surface to be dynamically smoother.
4. Distributions of data points from windsea tank experiments and from field experiments coincide with each other on the $C_D$ vs. $k_L/U_{10}$ and the $C_D$ vs. $k_L$ diagrams.

We propose further studies of the behavior of the nondimensional air–sea $CO_2$ transfer coefficient, as well as $C_D$, as a function of $R_B$, under various swell conditions, and for unsteady winds, in order to match the contemporary evolution of oceanic data sets with powerful new numerical simulations and ocean-wave models that are presently being generated.

## 1 Introduction

Coupled atmosphere–ocean models have made remarkable progress in recent years. In associated observational programs for global oceans, a new system of Argo floats are being deployed, which will collect 3000 CTD data points a day over the worlds' oceans. Moreover, enhanced satellite observations and a four-dimensional assimilation system, which the Global Data Assimilation Experiment (GODAE) will develop, are opening a new age of oceanography, a data revolution. In this new age, estimates of air–sea fluxes, including momentum and gas transfers, appear out of date, using only the 10-m wind speed and the air–sea partial pressure difference for gas transfers, respectively, neglecting the effect of sea state. With regard to the latter, Soloviev and Schlüssel [1] and others have proposed the necessity of sea-state dependence if gas transfer estimates are to be improved.

This chapter considers the challenge of determining a better parametrization to estimate the global $CO_2$ transfer between the atmosphere and the oceans, in light of laminar–turbulent regimes at the air–sea interface, which are greatly related to local wind and *wind waves*. Hereafter, we will use the term *windsea* for *wind waves*, after the terminology of Csanady [2], to express special fluid dynamical phenomena of wind waves, such as nonlinear coupling of water waves with local wind drift and turbulence, caused by the wind at the air–sea interface (cf. Toba [3, 4]). Regarding processes related to momentum and the gas transfers, both the skin friction and the form drag are important for the momentum transfer, whereas for the gas transfers the mechanism corresponding to the form drag is missing and contamination by small surface-active impurities is an important factor. However, in both cases the transfers are driven by turbulence rather than mere interface transfers. Wind-wave breaking, including incipient breaking, is closely related to the surface renewal as well as to the state of turbulence.

Toba and Koga [5] proposed the breaking-wave parameter $R_B$ to systematically describe various air–sea boundary processes. $R_B$ is defined by $u_*^2/\nu\sigma_p$, where $u_*$ is the air friction velocity, $\nu$ is the kinematic viscosity of air, and $\sigma_p$ is the windsea spectral peak frequency. In this chapter, we rename $R_B$ the *windsea Reynolds number*, since it is, from dimensional considerations, the only fundamental Reynolds number to describe fluid dynamical similarity conditions for *windsea boundary layer* phenomena, under the conditions of *wind–windsea equilibrium*. From observational data analyses, $R_B$ has already proven very effective in expressing the windsea breaking rate (Toba and Koga [5]), the whitecap coverage (e.g., Zhao and Toba [6]), and sea-spray droplet production (Iida *et al.* [7]). We will consider the behavior of the transfer coefficients for momentum and $CO_2$.

We briefly review the present status of the transfer coefficients: momentum transfer in Section 2.1 and $CO_2$ transfer velocity $k_L$ in Section 2.2. Then, in Section 3.1 we present a review of the wind–windsea equilibrium dynamics resulting in the similarity laws for windsea, the 3/2-power law, and in Section 3.2, using nondimensional analyses, we show that a single representative parameter can describe the windsea state itself, or the wave age. We also show the importance of $R_B$ in describing the associated fluid dynamical similarity for the air–sea boundary layer including pure windsea phenomena. We review the past data on availability of $R_B$ for wave-breaking related phenomena in Section 3.3. In Section 4 we present a new description of the behavior of the drag coefficient $C_D$ in light of $R_B$, and we provide motivation for its interpretation as the *windsea Reynolds number*. We will point out that the traditional transfer coefficients, as functions only of the wind speed and of the air–sea partial pressure difference for gases, are unstable quantities by themselves, because of dominating nonequilibrium conditions: dynamical contamination such as swells and the unsteady winds. In Section 5 we present new findings with respect to the similarity between the behavior of $C_D$ and $k_L$ from the viewpoint of $R_B$, using observational data. We give physical interpretations, as far as possible, to propose a new *nondimensional $CO_2$ transfer velocity*.

Finally, we propose further studies of the nondimensional transfer coefficients, to provide better estimates of global air–sea transfers appropriate for this age of the ocean data revolution. We also briefly discuss the outstanding questions that remain to be investigated.

## 2 Present status of air–sea transfer coefficients

### 2.1 Momentum transfer (drag) coefficient

Historically, the sea-surface wind stress, $\tau$, was expressed by

$$\tau = \rho C_D U_{10}^2 \ (=\rho u_*^2), \tag{1}$$

where $\rho$ is the density of the air, $C_D$ is a dimensionless drag coefficient, and $U_{10}$ is the 10-m wind speed. This equation actually defines $C_D$ and the air friction velocity $u_*$.

The original meaning of eqn (1) is

$$\tau = C_D \Delta M U_{10}, \quad \Delta M = \rho U_{10}, \tag{2}$$

where $\Delta M$ is the difference of air momentum between the 10-m level and the sea surface, and it is assumed that $\tau$ should be proportional to both $\Delta M$ and $U_{10}$. For a solid surface, $C_D$ is actually a constant of the order of $10^3$, differing according to the nature of the surfaces. Windsea exists at the air–sea boundary, and observations have shown that $C_D$ is not a constant but highly variable, even under neutral stratification conditions. It has traditionally been assumed that this scatter may be expressed as a function of $U_{10}$. However, $C_D$ should actually be a measure that extracts (or amplifies) delicate variations due to the unsteady wind and the state of windsea, though observational errors are also being amplified.

A series of recent observations have shown a compact distribution. An example is a formula presented by Yelland *et al.* [8], expressed as

$$10^3 C_{DN} = (0.50 \pm 0.02) + (0.071 \pm 0.002) U_{10N} \quad (6 < U_{10N} < 26\,\mathrm{m/s}), \tag{3}$$

where the subscript N indicates neutral stratification. This data was obtained by the inertial dissipation method in deep water, with a correlation coefficient of 0.80 for 1111 data points. However, there was considerable scatter among individual measured values (e.g., Yelland and Taylor [9], their fig. 7.5). This scatter was considered to be caused by the effect of windsea, in seas ranging from young to old, and swells.

There have been many attempts to determine the cause of the data scatter, but they have not been fully successful. Figure 1 shows one of these recent efforts. It is a synthesis of observational data from laboratory experiments as well as open ocean data, with an overall parametrization proposed by the SCOR Working Group 101 (Jones and Toba [10]). The data are expressed in terms of the nondimensional roughness parameter ($gz_0/u_*^2$) as a function of the inverse wave age $\beta^{-1} = (u_*/C_p)$, where $g$ is the acceleration due to gravity, $z_0$ is the aerodynamic roughness parameter (associated with the neutral stratification $C_D$), and $C_p$ is the phase speed of the wind-wave peak frequency. The latter represents the state of windsea conditions. The SCOR formula was expressed as

$$\begin{aligned} gz_0/u_*^2 &= 0.03\beta \exp(-0.14\beta), \quad \sim 0.35 < \beta < 35 \\ &= 0.008, \quad \beta \geq 35 \text{ (light winds over swell).} \end{aligned} \tag{4}$$

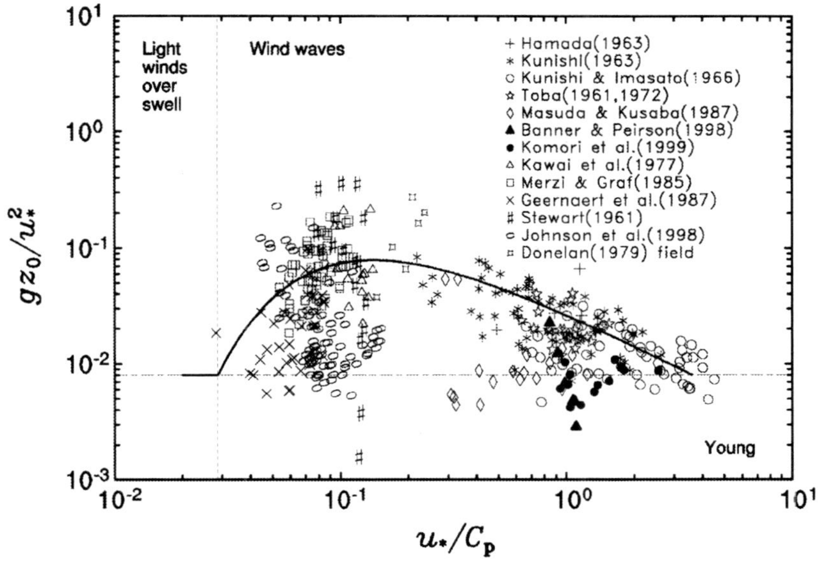


Figure 1: A synthesis of sea-surface drag data observed at laboratories and seas. The data are expressed in terms of the nondimensional roughness parameter ($gz_0/u_*^2$) as a function of the inverse wave age $\beta^{-1} = (u_*/C_\mathrm{p})$. An overall formula [eqn (4)] proposed by the SCOR Working Group 101 is shown as a thick curve. The figure is from Jones and Toba [10] with some additional recent data. The right hand cluster of data points are from laboratory experiments and the left hand cluster are from field observations.

The Jones and Toba [10] SCOR report presented comprehensive reviews on various effects that can cause the scattering in measurements of $C_\mathrm{D}$. These included the presence of windsea and swell, unsteady winds, atmospheric boundary layer rolls, large time and space averages of data, surface tension biases in observations, spatial inhomogeneity, basin boundaries, and the effect of stability, constituting another important atmospheric boundary layer factor. Recent data taken at a tower station by Suzuki *et al.* [11] showed that crosswind swells also significantly reduced $C_\mathrm{D}$.

## 2.2 $CO_2$ transfer velocity

The $CO_2$ flux $F$ through the water surface has usually been expressed by:

$$F = k_\mathrm{L} S \Delta p\mathrm{CO}_2, \tag{5}$$

where $k_\mathrm{L}$ is the gas exchange coefficient, $S$ is the solubility of the gas, and $\Delta p\mathrm{CO}_2$ is the difference between the partial pressures of the gas in air and in water. The value of $k_\mathrm{L}$ is to be determined empirically. It is important that $k_\mathrm{L}$ has the dimensions of speed (m s$^{-1}$), which is different from the case of $C_\mathrm{D}$.

Although the $CO_2$ transfer velocity $k_L$ across the air–sea interface has been measured by a number of investigators in oceans and lakes (McGillis *et al.* [12]; Fairall *et al.* [13]; Haubecker and Jahne [14]; Wanninkhof *et al.* [15]; Watson *et al.* [16]; Cember [17]; Broecker *et al.* [18]), the field data for $k_L$ plotted against $U_{10}$ are very scattered among individual studies. The scatter may have been caused by factors such as windsea, swell, density stratification, surfactants, bubbles, and so on (e.g., Komori and Shimada [19]; Komori and Misumi [20]).

However, one of the biggest problems in previous studies was that the scattered $k_L$ data were usually correlated by monotonous increasing curves with $U_{10}$ (e.g., Liss and Merlivat [21]; Wanninkhof [22]) and the validity of the correlation with $U_{10}$ has not been physically discussed.

The $CO_2$ transfer across the air–sea interface is controlled by a laminar–turbulent regime shift of the windsea boundary. Turbulent motions on the liquid side beneath the interface are especially important. Komori *et al.* [23] first verified that surface-renewal eddies on the water side control heat transfer across the air–water interface. The significant role of surface-renewal eddies on the $CO_2$ transfer has been confirmed in both an open channel flow with an unsheared air–water interface (Komori *et al.* [24]) and a wind-driven turbulence with a sheared interface (Komori *et al.* [25]). These studies have found that $k_L$ follows the surface-renewal concept in the low and middle wind speed region without wave breaking. That is, $k_L$ is proportional to the square root of the frequency of appearance of surface-renewal eddies $f_s^{1/2}$. In addition, Komori *et al.* [26] showed that $k_L$ remarkably increases in the wave-breaking region, since the effect of very tiny surfactants disappears owing to wave breaking. These results suggest that $k_L$ should be correlated not solely with the wind speed but with a parameter that should be constituted by the surface-renewal frequency $f_s$, and also the wave-breaking ratio. However, it is almost impossible to estimate the surface-renewal frequency directly from available remotely sensed observational data, and therefore we have to find a new parameter that reflects both the turbulence structure and the wave-breaking occurrence.

A recent paper by Zhao *et al.* [27] is one effort in this direction. We will present a further comprehensive effort in Section 5.

# 3 Wind–windsea equilibrium and nondimensional windsea boundary layer variables

## 3.1 Wind–windsea equilibrium: similarity laws of windsea

Nonlinear interactions between the wind and the windsea are very strong. They include turbulence, shear flows, the form drag and breaking of the air–water interface, and the related microphysical processes at the air–water interface, together with the breaking of waves themselves. As a result of these strong interactions, there are statistical similarity laws particular to the windsea field. In this section, these similarity laws are briefly reviewed. The similarity laws give a basis for the pure windsea field being expressed by using a single nondimensional parameter,

and also for the importance of the windsea Reynolds number. A detailed review of these similarity laws is given by Toba [3, 4, 28].

### 3.1.1 The 3/2-power law

For growing windsea under steady wind conditions, i.e., for windsea in local equilibrium with the wind, there exists the 3/2-power law which is expressed as:

$$H^* = BT^{*3/2}, \quad B = 0.062, \tag{6}$$

where $H^* = gH_s/u_*^2$ and $T^* = gT_s/u_*$, $H_s$ is the significant wave height, $T_s$ is the significant wave period, and $B$ is an empirical constant. Alternatively, eqn (6) may be expressed in a dimensional form:

$$H_s = B(gu_*)^{1/2}T_s^{3/2}. \tag{7}$$

Equation (6) compares well with data as seen from fig. 2–6 of Toba [28], or fig. 23 of Toba [4]. Ebuchi *et al.* [29] showed that the 3/2-power law with the above value of $B$ holds even for cases with some swells, though not under dominant swells (see also Johnson *et al.* [30]).

This power law was originally derived by Toba [31, 32] by eliminating nondimensional fetch from empirical formulas for the growth of wind waves, e.g., as given by Wilson [33], Mitsuyasu *et al.* [34], and the JONSWAP formula by Hasselmann *et al.* [35]. However, the 3/2-power law is not a consequence of fetch laws for each of $H_s$ and $T_s$, but of the local equilibrium between the wind and windsea, where $H_s$ and $T_s$ are intrinsically related to $u_*$.

As a matter of fact, it was already shown that eqn (6) is just equivalent to the condition of proportionality

$$u_* \propto (\overline{u_a'^2})^{1/2} \propto (\overline{u_w'^2})^{1/2} \propto u_{*w} \propto u_0, \tag{8}$$

where $(\overline{u_a'^2})^{1/2}$ is the intensity of air turbulence, $(\overline{u_w'^2})^{1/2}$ is the intensity of water turbulence, $u_{*w}$ is the water friction velocity, and $u_0$ is the Stokes drift velocity (see Toba [3, 28, 32]).

Fetch is usually difficult to define in the oceans, since the winds are always varying. Nevertheless the 3/2-power law holds, especially for strong wind cases, or for cases free of dominant swells (Ebuchi *et al.* [29]). Also, Wilson's fetch graph formulae [33] show that the 3/2-power law holds for conditions approaching long-fetch saturation, where the empirical fetch laws for the nondimensional $H_s$ and $T_s$ have complicated forms.

In the sea, the value of $B$ was estimated as $0.062 \pm 20\%$ by Toba *et al.* [36] as derived from measured data. The scatter is largely due to unsteady winds, namely conditions where windsea is not in local equilibrium with the wind, and the windsea is over- or undersaturated (Toba *et al.* [37, 38]; Waseda *et al.* [39]). The influence of unsteady winds on the wind stress was reviewed in chapter 9 of Jones and Toba [10]. Komatsu and Masuda [40] numerically estimated Toba's constant $B$ to have a "standard" value of 0.06, using the RIAM method to numerically evaluate nonlinear energy transfers among wind waves, as well as by using the WAM method (The WAMDI Group [41]). Suzuki [42] reported some similar results by using his JWA3G wave model.

Alternatively, the 3/2-power law can be regarded as the situation where the steepness of significant waves is statistically limited, as a function of the wave age of significant waves $\beta = C_s/u_*$, as an equilibrium state of windsea under the action of the local wind. This relation is expressed by

$$\frac{H_s k_s}{2\pi} = (2\pi)^{1/2} B \left(\frac{C_s}{u_*}\right)^{-1/2} \left\{1 + u_s^* \left(\frac{C_s}{u_*}\right)^{-1}\right\}^{-2/3}, \quad u_s^* \equiv \frac{u_s}{u_*} = 0.206, \quad (9)$$

where $k_s$ is the significant wave number, $C_s$ the phase speed of significant waves, and $u_*$ is the friction velocity of air (Bailey *et al.* [43]). The wave age in the second term is the consequence of an averaged local wind drift $u_s$ (Tokuda and Toba [44]).

### 3.1.2 One-dimensional windsea frequency spectral form

In observed windsea, there is an equilibrium range in the main part of the high frequency side of the energy spectrum. The accepted one-dimensional windsea frequency spectrum for this range, proposed by Toba [45] from observational data with some dimensional considerations, is given by

$$\phi(\sigma) = \alpha_s g u_* \sigma^{-4}, \quad \sigma > \sigma_p, \quad (10)$$

where $\alpha_s$ is a constant, $\sigma$ is the angular frequency, and $\sigma_p$ is the peak angular frequency of windsea. This spectral form has been shown to be consistent with many observational data sets since Kawai *et al.* [46] and Phillips [47] reproposed this form theoretically. Examples of observed data are seen in fig. 2–8 of Toba [28].

Zakharov and Filonenko [48] predicted theoretically that the one-dimensional spectral form should be proportional to $\sigma^{-4}$ for surface gravity wave interactions. However, there seems to be no study which derives the $u_*$-proportionality purely from theory. The $u_*$-proportionality is the most important observational fact for windsea, and the 3/2-power law corresponds to the integral form of this spectral form.

The value of the coefficient $\alpha_s$ seems to be in the range $6 - 12 \times 10^{-2}$ (e.g., Phillips [47]). Joseph *et al.* [49] suggest that $\alpha_s$ should be 0.096 for consistency with the 3/2-power law with $B = 0.062$, for self-similar spectra. However, $\alpha_s$ shows variability in response to the wind fluctuations, just as in the case of $B$ values, corresponding to the over- and undersaturation of the wave energy (Toba *et al.* [38]).

### 3.1.3 Physical implications

The similarity laws are conceptually expressed, for the 3/2-power law, as

$$f_1(H_s, T_s, u_*) = 0, \quad (11)$$

and for the energy spectrum as

$$f_2(\sigma_p, \phi_p, u_*) = 0, \quad (12)$$

where $\phi_p$ is the spectral density at $\sigma_p$ and, as a matter of fact, eqn (11) represents the integral form of eqn (12). In any case, specification of two variables, e.g., $u_*$ as the external forcing field and another variable, determines the state of windsea, assuming it is free of dominant swell. As for the physical background of the similarity laws, they are results of a series of wind-forced, strongly nonlinear processes. It is important that the external condition is $u_*$, and there is $u_*$-proportionality in the windsea. These laws are the consequence of self-adjustment processes in windsea forced by $u_*$.

It is noteworthy that Tulin [50] suggested that the 2/3-power law can be explained by a simultaneous consideration of wave energy and wave momentum change rates resulting from wave breaking, keeping the water wave relation:

$$M = E/C, \tag{13}$$

where $M$ is the wave momentum, $E$ is the wave energy, and $C$ is the phase speed. This leads to the necessary downshifting of wave frequency, to satisfy the 2/3-power law, keeping a balance between the wind input and dissipation by wave breaking. Further review of the processes of the local equilibrium is given by Toba [4], Csanady [2], Waseda *et al.* [39], and Toba [28].

The existence of the wind–windsea equilibrium, or the similarity laws, as described in this section provides a basis for our analyses, using the windsea Reynolds number (breaking-wave parameter).

### 3.2 Fundamental dimensional considerations

The usefulness of nondimensionalizing variables for quantitative analyses of complicated experimental data was comprehensively described in Section 2–1 of Toba [28], together with its practical procedures and applications. By nondimensionalization the number of variables can be reduced by the number of fundamental dimensions originally involved (the Pi theorem), and we can abstract essential characteristics related to the respective phenomenon. The method of nondimensionalization is especially useful for processes where turbulence dominates. In other words, even when experimental data in dimensional quantities are so complex that it is almost impossible to find any law from them, it becomes easy to extract essential relationships or laws by reducing variables in dimensionless forms. However, formal application of the Pi theorem for solely diminishing variables does not tell us how to combine the variables to make nondimensional products, or how to select mutually independent nondimensional variables. In order to construct meaningful nondimensional variables, we must invoke physical insight.

In the case of windsea, the *independent variables* are time and space, or duration and fetch if the wind field is homogeneous. The *dependent variables* are wave property variables such as $H_s$, $T_s$, or $\sigma_p$, $\phi_p$, and $u_*$ as the only *external condition* (or *conditioning variable*), at least under neutral stratification and in the absence of dominant swells. The existence of the similarity laws described in Section 3.1 implies that it is sufficient to select *only one* of these wave-property variables, together with $u_*$, in order to completely describe the dynamical system.

In terms of *physical constants*, we have the acceleration due to gravity $g$ and the kinematic viscosity of air $\nu$. The viscosity is substantial since the air–sea boundary processes are basically those of a viscous turbulent fluid, and $g$ appears explicitly in eqn (10). Since the kinematic viscosity of water $\nu_w$ has a specific ratio with $\nu$, we need consider only $\nu$. We do not need to consider the surface tension, since it is only related to very high frequency waves, and that part of the windsea spectra is actually controlled by $u_*$, e.g., proportional to $u_*^{2.5}$ (Mitsuyasu and Honda [51]), forming the tail portion of the continuous windsea spectrum.

There are five quantities with two fundamental dimensions of time and length, and we have one relationship [eqn (11) or eqn (6)]. Thus, it is sufficient to construct two fundamental nondimensional variables:

$$H^* = gH_s/u_*^2 \quad \text{or} \quad T^* = gT_s/u_* \quad \text{using eqn (6)}, \tag{14}$$

$$R_B = u_*^2/\nu\sigma_p \quad \text{or} \quad u_*^2T_s/\nu \quad \text{or} \quad u_*H_s/\nu \qquad \text{using eqns (6) and (12)}. \tag{15}$$

Here, $H^*$ is interpreted physically as a *measure of the windsea development for a given wind* $u_*$, since it contains $g$ and $H_s$ (or $T_s$) in the numerator and $u_*$ in the denominator. The variable $R_B$ is interpreted as a *measure of the fluid dynamical conditions at the air–water boundary*, since $u_*$ and $T_s$ (or inverse $\sigma_p$), or $H_s$, enters both in the numerator. We call this the *windsea Reynolds number*.

As a matter of fact, $R_B$ may be interpreted as a Reynolds number, since it contains $u_*$ as a representative scale of speed, and $L_s = u_*T_s(=2\pi u_*/\sigma_p)$ as a representative length scale of the phenomenon under consideration. $L_s$ is interpreted as the length representing the distance over which a water particle at the surface is driven by the speed $u_*$ (representing the wind stress) during the representative wave period $T_s$. For the same value of a Reynolds number $Re$ ($=UL/\nu$ with $U$ and $L$ as the representative scales of speed and length), the equation of motion will have the same solution, even if $L$, $U$, and/or $\nu$ have different values. This is known as the *Reynolds' similarity law*, which provides the basis for model experiments.

As will be described in Section 3.3, $R_B$ was proposed by Toba and Koga [5] to properly express various air–sea boundary processes including the wind stress, wave breaking, bubble entrainment, and generation of sea-spray droplets, and later Toba *et al.* [52], Zhao and Toba [6], and Toba [28] denoted this variable as the *breaking-wave parameter*. $R_B$ is expected to be closely related to $C_D$ and $k_L$, as will be discussed later in Section 4. Moreover, it should be noted that $T^*$, which is related with $H^*$ by the 3/2-power law (6), may be regarded as the *wave age of* the *significant waves*, $\beta = C_s/u_*$, as was earlier pointed out by Toba [31].

## 3.3 Windsea Reynolds number: fundamental parameter of the windsea boundary layer

To quantify wave breaking, photographs of the sea surface were used to determine the percentage area of whitecap coverage. Most researchers treated it only as a function of the wind: $U_{10}$ or $u_*$ (e.g., Monahan and MacNiocaill [53]; Wu [54]).

In view of the situation that windsea is in local equilibrium with the wind, or has the similarity laws discussed in the previous sections, windsea parameters in addition to wind should be used to describe phenomena related to the air–sea interface. We should invoke the nondimensional analyses as discussed in Section 3.2. Actually, this was done with $R_B$ proposed by Toba and Koga [5]. Figure 2(a) shows the percentage $\alpha$ of breaking crests, among individual waves, passing a fixed point on the water surface, plotted against $U_{10}$. It is notable that data from short-fetch, wind-wave tank experiments systematically differ from those obtained at a tower station at sea. This shows that $\alpha$ greatly depends on the fetch, which indicates dependence on the degree of windsea development. However, if we plot $\alpha$ against $R_B$, all the data collapse along a line with an angle of 45°, as shown in Fig. 2(b). We note that $\alpha$ was originally a nondimensional quantity.

Toba and Koga [5] also successfully applied $R_B$ to describe the whitecap coverage, and the sea-spray droplet concentration observed near the sea surface.

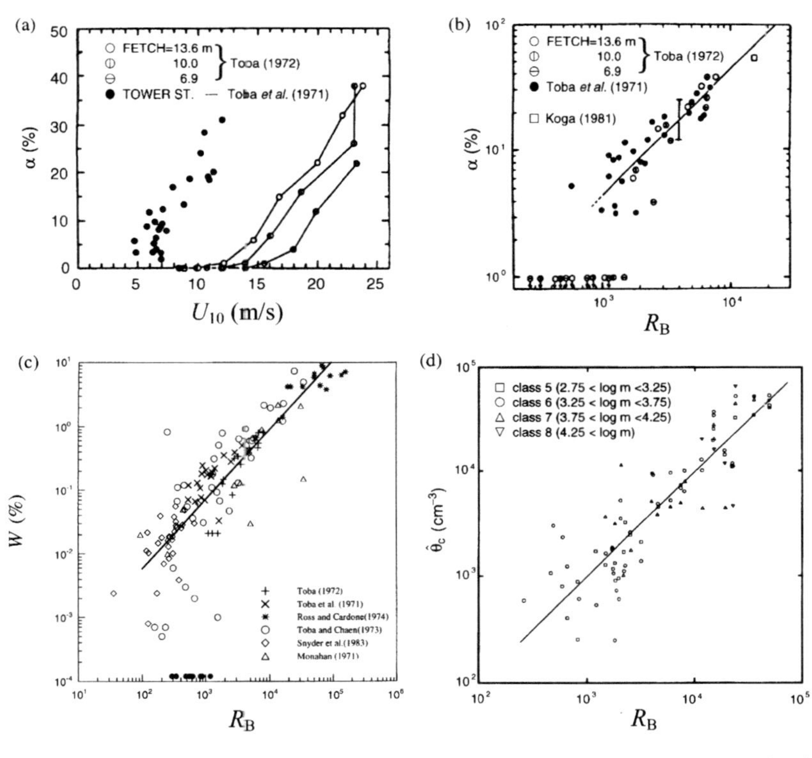


Figure 2: (a) Percentage $\alpha$ of breaking crests among individual waves of windsea passing a fixed point on the water surface, plotted against $U_{10}$, and (b) plotted against $R_B$ (from Toba and Koga [5]). (c) Whitecap coverage percentage, $W$, plotted against $R_B$ (from Zhao and Toba [6]). (d) Seawater droplet concentration observed near the sea surface, plotted against $R_B$ (from Iida et al. [7]).

Figure 2(c), from Zhao and Toba [6] with additional data sets, shows that all the data collapses along the 45°-line, and the correlation coefficient is actually 0.88. Wave breaking starts to occur at around $R_B = 10^3$, as suggested by Toba and Koga [5]. Zhao and Toba [6] showed that $R_B$ is the most effective variable, since other variables had correlation coefficients much smaller than $R_B$, e.g., 0.43 using wave age, and in the cases for $H_s$, $T_s$, and $u_*$, correlation coefficients were not larger than 0.80.

Zhao and Toba [6] also reported that another Reynolds number, $R_H = u_* H_s/\nu$, had a correlation coefficient 0.84. Although this is smaller than 0.88, the variation in the whitecap coverage is similar to $R_B$, since it is also a Reynolds number representing the same physical processes. In fact, Toba and Kawamura [55] reported that the depth of the associated windsea boundary layer [*downward bursting boundary layer* (*DBBL*)] is 5 times $H_s$. The concept of this Reynolds number was first used by Toba and Kunishi [56], and later by Banner and Peirson [57]. In using $R_B$, determination of $\sigma_p$ may be achieved by taking the peak frequency of the windsea part of the spectra using the methodology proposed by Hanson and Phillips [58]. However, for $R_H$ computations, $H_s$ may be affected by the ever-present existence of swells. Thus $R_B$ is better than $R_H$ as far as wave spectral data are available.

The sea-spray droplet concentration observed near the sea surface also shows a similar distribution, along the 45°-line, as shown in Fig. 2(d) (from Iida *et al.* [7]). Thus the windsea Reynolds number $R_B$ can be considered a crucial parameter for describing macroscopic windsea boundary layer phenomena: windsea breaking, air entrainment, whitecap coverage, and sea-spray droplet (sea-salt particles) production.

Toba and Koga [5] also pointed out the importance of $R_B$ in specifying the sea-surface wind stress. However, we will revisit this matter with a new series of data, in Section 4.

## 4 Drag coefficient in light of the windsea Reynolds number

As discussed in Section 3, $R_B$ is expected to be a crucial parameter in describing the behavior of nondimensional air–sea transfer coefficients. In this section, we examine it for $C_D$, and then apply the concept to the $CO_2$ exchange questions in Section 5.

### 4.1 Laboratory data: laminar–turbulent boundary layer transition

Figure 3 shows $C_D$ plotted against $R_B$, as measured by wind-wave tank experiments in fetch-limited (stationary) conditions. The same data set is plotted against $U_{10}$ and against $u_*$ in Fig. 4a and b. These data correspond to conditions where *windsea* is *in local equilibrium with the wind*. These data are the cluster of data points with larger values of $u_*/C_p$ in Fig. 1. The scatter of the data is apparently much smaller in Fig. 3, when expressed as a function of $R_B$, than when expressed against $U_{10}$ or $u_*$ in Fig. 4a and b. Note that, in the case of Fig. 4b, $u_*$ already includes $C_D$, after eqn (1). That Fig. 3 shows remarkably better correlation than in Fig. 1 indicates that the wave age is not a proper parameter to discriminate detailed momentum transfer phenomena.

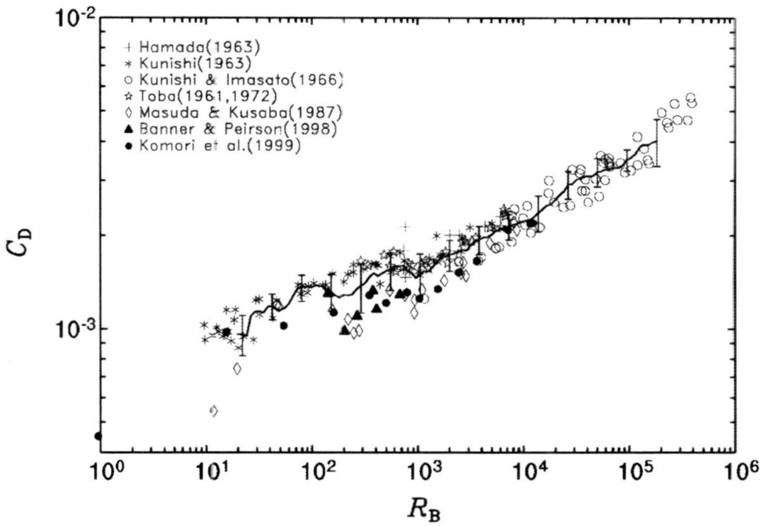


Figure 3: The drag coefficient $C_D$ measured by wind-wave tank experiments in fetch-limited (stationary) conditions plotted against the *windsea Reynolds number* $R_B$.

An important finding from Fig. 3 is that we can see two critical points in the distribution of $C_D$, at $R_B = 2 \times 10^2$ and at $R_B = 10^3$. At these points, the data indicate two downward points, which bracket a bulge region. Beyond $R_B = 10^3$, the $C_D$ values become very large. Figure 4a and b does not show any such critical points in a clear-cut way as shown in Fig. 3. The former critical point, $R_B = 2 \times 10^2$, may be regarded as the transition from a smooth surface to the occurrence of airflow separation, and the creation of the form drag. The latter conspicuous critical point, $R_B = 10^3$, corresponds to the transition to the fully turbulent regime of the windsea boundary layer with the occurrence of windsea breaking, as stated in Section 3.3.

As a matter of fact, by using flow visualization techniques Kawai [59, 60] and Kawamura and Toba [61] gave detailed descriptions of the airflow structure over windsea surfaces in a wind-wave tank. Kawamura and Toba [61] demonstrated the existence of a high shear layer at the outer edge of separation bubbles behind laboratory wind-wave crests (their figs 22 and 23). Ebuchi *et al.* [62] described changes of the windsea surfaces from smooth to rough by using an optical method (their fig. 4). Readers can also refer to photographs showing the change of the windsea surfaces in Toba [63], table in Toba [31], and Koga [64]. The conditions of the windsea surfaces, together with the nature of airflow over water, change with the development of windsea as well as with increasing wind speed. Thus, the overall situation should be specified by the *windsea Reynolds number*, $R_B$.

Regime shifts of these conditions as a function of $R_B$ are shown schematically in Fig. 5. The top panel shows the conditions where the windsea surface is smooth and $R_B$ is small, corresponding to the leftmost part of Fig. 3 ($R_B$ smaller than $10^1$). The second panel in Fig. 5 illustrates the conditions for the presence of a separation

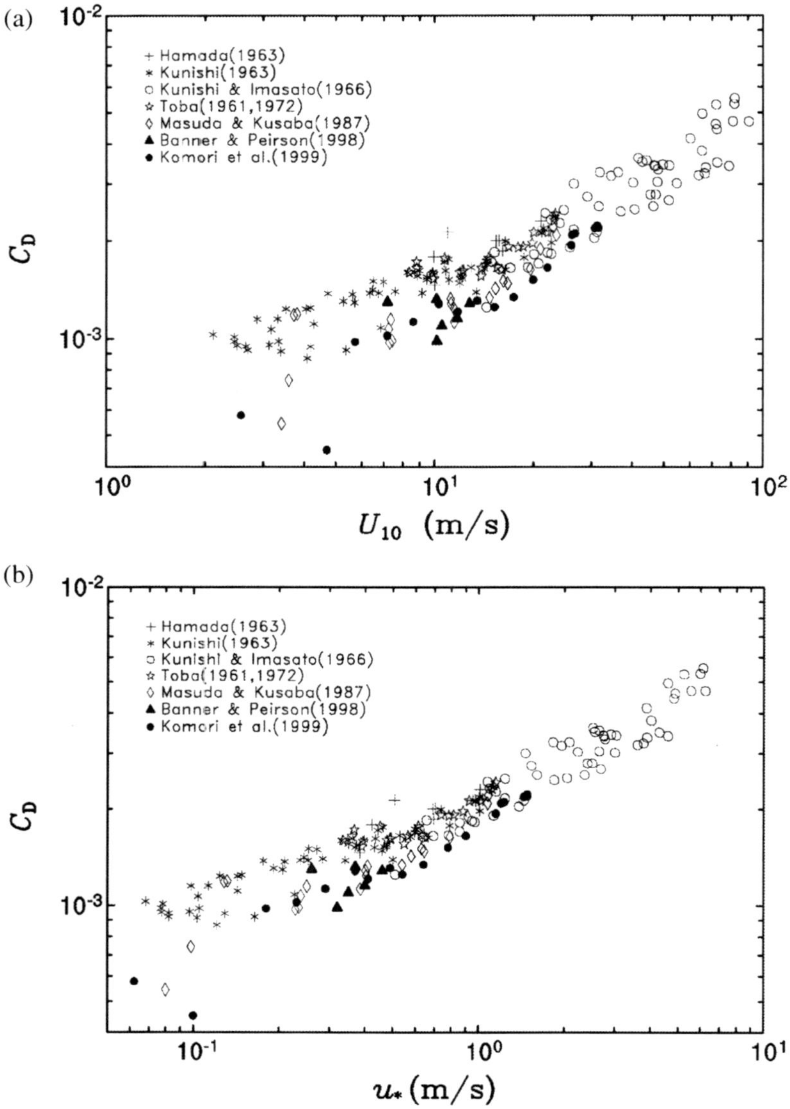


Figure 4: The same data as in Fig. 3 for the drag coefficient $C_D$ measured by wind-wave tank experiments in fetch-limited (stationary) conditions, but plotted against (a) $U_{10}$ and (b) $u_*$.

point just at the lee side of the crest (about $3 \times 10^2$ in $R_B$), corresponding to figs 22 and 23 of Kawamura and Toba [61]. Under the crest, a water recirculation region (black area in the figure) exists (cf. Okuda [65]).

As $R_B$ increases, a streaky structure appears, then the windsea surface becomes rough, as demonstrated by Ebuchi *et al.* [62]. In analogy to Shapiro [66],

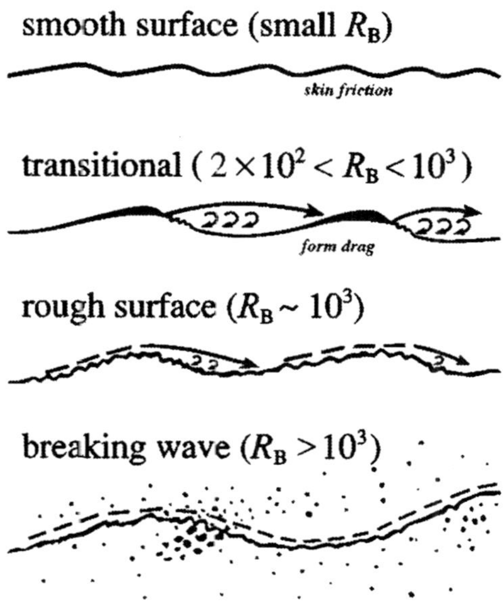


Figure 5: Schematic representation of the regime shift of the windsea boundary layer conditions as a function of $R_B$.

we hypothesize that the separation point would shift to the windward face of the crest, and the turbulent boundary layer would become thinner with a smaller separation bubble, as illustrated in the third panel of Fig. 5 (about $10^3$ in $R_B$). At this rough-surface stage, $C_D$ would even decrease as shown in Fig. 3 for $R_B$ approaching $10^3$. However, because the windsea is composed of various irregular individual wind waves, a sharp decrease in $C_D$, as shown in the analogous case for a system of golf balls, does not occur and the curve in Fig. 3 is flatter with a slight kink at $10^3$ in $R_B$.

When $R_B$ exceeds $10^3$, wave breaking commences as shown by Toba and Koga [5] and Zhao and Toba [6], turbulence breaks the interface, and wind stress, as well as gas transfer, increases drastically, corresponding to the conditions on the right side of Fig. 3. The bottom panel of Fig. 5 illustrates the conditions, $R_B \sim 5 \times 10^3$ or so, corresponding to photographs shown by Koga [64, 67].

As discussed in Section 3, windsea is a special fluid dynamical phenomenon at the air–water interface, where the coupling process between the two boundary layers of air and water is important, and includes elements of the local wind drift, turbulence in the air and water boundary layers, together with the surface wave motion (Toba [3, 4]; Csanady [2]). The wind stress (surface drag) is related to all these processes, including flow separation, pressure distribution, and viscous stress. Consequently, it is very reasonable that the momentum transfer is controlled by the windsea Reynolds number $R_B$.

Figure 6 shows laboratory observations of tangential stress and total stress (including form drag) by Banner and Peirson [57]. The values of $R_B$ shown in the abscissa were calculated by using their observational data including windsea, and

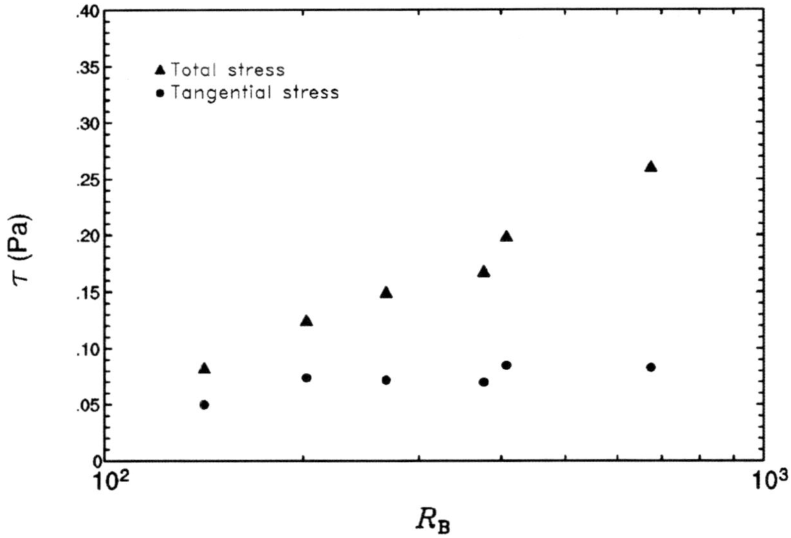


Figure 6: Laboratory observational data of tangential stress and total stress (including form drag) from Banner and Peirson [57]. The values of $R_B$ shown on the abscissa were calculated using their reported data including windsea.

it fortunately ranges between $10^2$ and $10^3$ in $R_B$. It is seen that tangential stress stops increasing at $2 \times 10^2$, whereas total stress continues to increase. This corresponds to our inference that the shift from the skin friction regime to the form drag regime should occur between the above two critical points.

### 4.2 Field data: shift of critical point with dynamical contamination

Figure 7 repeats Fig. 3, adding the same data as in Fig. 1, in particular the cluster of points located at the smaller values of $u_*/C_p$ for field observations. The above-mentioned critical point of $10^3$ in $R_B$, seen in Fig. 3, seems to shift to larger values in $R_B$ in the field data in Fig. 7, with $C_D$ tending to be much smaller. $C_D$ thus seems to be an *unstable* quantity that easily changes its character, in the presence of dynamic contamination, such as swells and unsteady winds (see, also Toba *et al.* [38]; chapter 9 of Jones and Toba [10]). Aside from observational errors, this should be the main reason why field observations show so much scatter, and frequently give relatively small values. Figure 8a and b shows the same $C_D$ data expressed against $U_{10}$ or $u_*$. Note that the data points exhibit much less systematic scatter than appears in Fig. 7.

Figure 7 reminds us of a schematic figure by Shapiro used to explain why golf balls have rough surfaces. Figure 9 is cited from fig. 92 of Shapiro [66]. Shapiro suggested that when the surface of a sphere located in an airflow is smooth, drag first increases with increasing flow speed while the boundary layer is laminar, and there is no flow separation. Flow separation then occurs just behind the shoulder at the

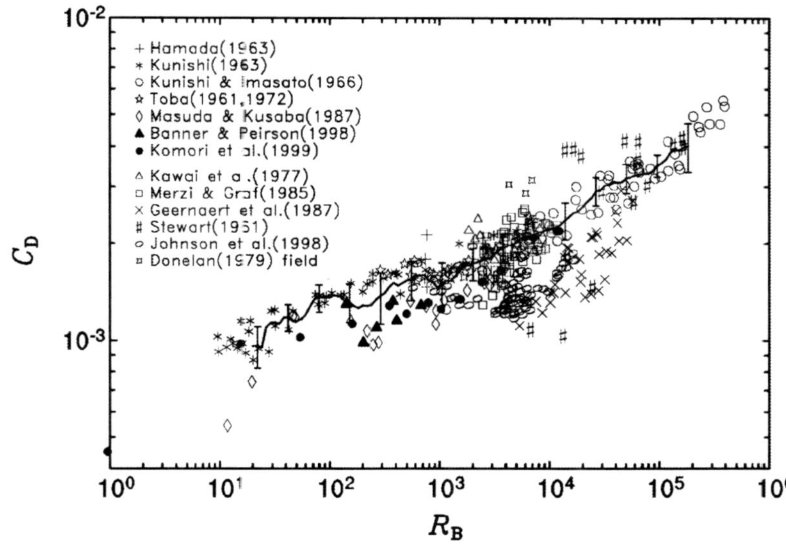


Figure 7: The same data as in Fig. 3, with the addition of the drag coefficient data in Fig. 1 for the cluster of points located at the smaller values of $u_*/C_p$, corresponding to field observations.

corresponding flow speed. The regime then shifts from laminar to turbulent conditions, and the drag decreases and jumps to another curve for the turbulent boundary layer. At this transition, a thick turbulent boundary layer appears downstream of the separation point. As the flow speed further increases, drag then increases again due to the generation of a wake (see the thin line in Fig. 9).

When the surface of a sphere is rough, the drag is initially larger than that for a smooth surface, because of the form drag due to the roughness. However, the occurrence of flow separation shifts to a lower flow speed, with the separation point shifting upstream of the shoulder, for a thinner turbulent boundary layer. This situation is just the same as illustrated by the third panel in Fig. 5. Though the drag increases again with increasing flow speed, there is a flow speed region (from A to B in Fig. 9) where the drag is smaller for the rough surface case. This region represents the typical speeds at which golf balls travel. The above explanation of Fig. 9 was given as a function of the flow speed in the frame of reference of the moving sphere. However, since golf balls have the same diameter, Fig. 9 can be similarly redrawn with the Reynolds number as the abscissa. This explanation can be extended to the case of the actual sea surface, as discussed above.

For conditions under steady wind without a notable swell component, growing windsea is in local equilibrium with the wind, and the 3/2-power law holds as described in Section 3 (the *wind–windsea equilibrium*). Under unsteady conditions of changing wind speed and direction, the windsea is continuously in the process of adjustment toward the local equilibrium with the ever-changing new wind. For example, an increasing wind causes undersaturation of the high-frequency side

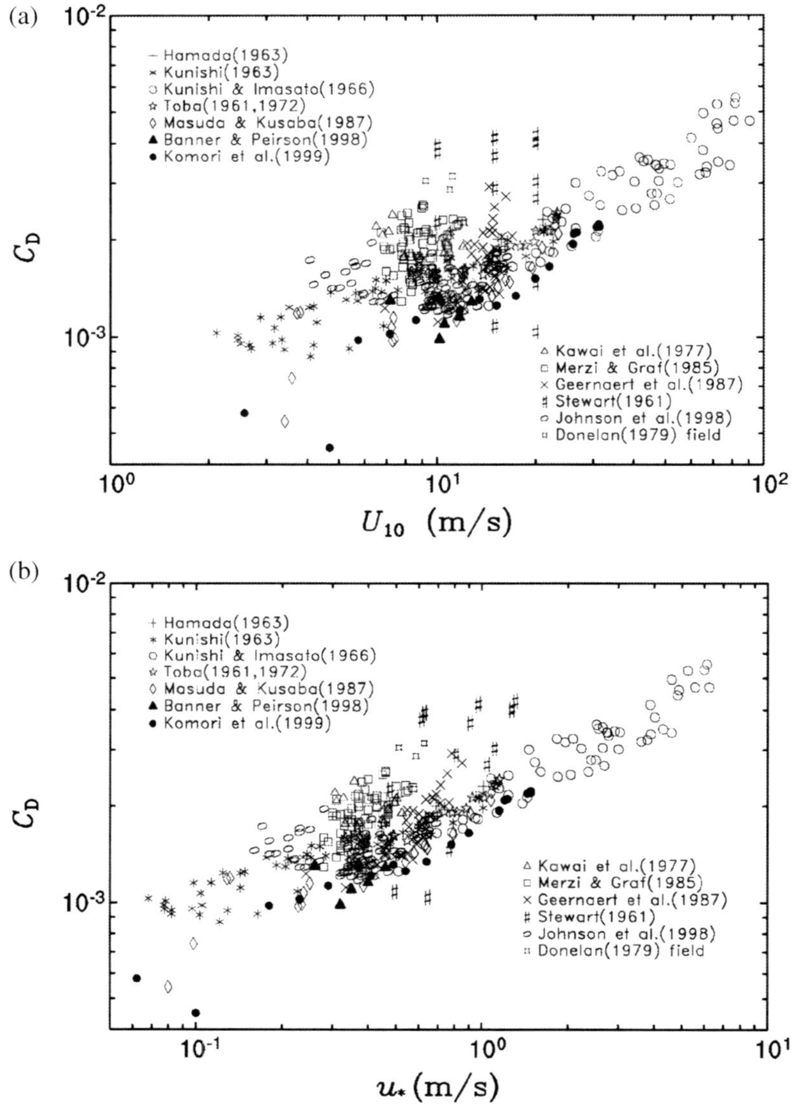


Figure 8: The same data as in Fig. 7 for the drag coefficient, but plotted against (a) $U_{10}$ and (b) $u_*$.

energy level of the windsea spectrum. Decreasing wind causes oversaturation (Toba *et al.* [37]; Hanson and Phillips [58]). These self-adjustment processes should manifest themselves as amplified fluctuations in $C_D$. The fluctuations of $u_*$ and $C_D$ thus correspond to under- and oversaturation phenomena (Toba *et al.* [38]).

The existence of conspicuous swell components causes inefficiency in the self-adjustment of windsea to the wind, resulting in smaller values of $C_D$, as

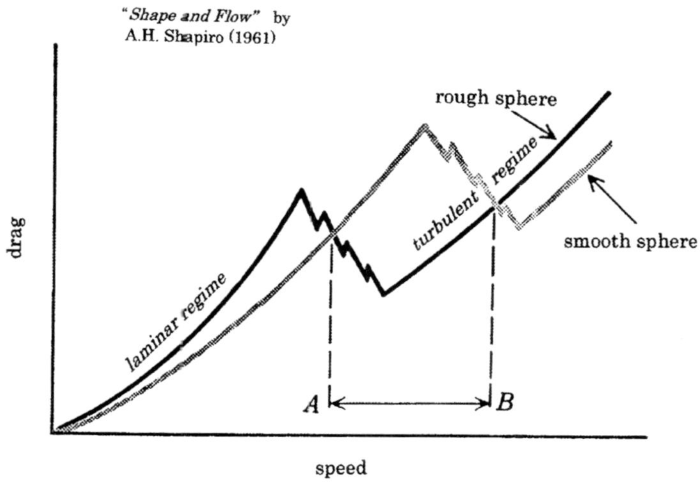


Figure 9: Schematic figure explaining why golf balls have rough surfaces, cited from fig. 92 of Shapiro [66].

demonstrated in Fig. 7. In fact, the existence of swells can cause the wind-wave energy to decrease compared with the pure windsea cases, in a manner reported by Hatori *et al.* [68] and Toba *et al.* [69] (see fig. 22 of Toba [4]). The sea surface would therefore be smoother, in a sense, in the presence of swells, as also reported by Taylor and Yelland [70]. Moreover, Suzuki *et al.* [11] presented the observational data showing reduction in the roughness length $z_0$ for the same wind, under the existence of swell components in a direction different from the local wind.

$C_D$ itself is thus considered an *unstable quantity*, and nonequilibrium conditions caused by unsteady winds or the existence of significant swells easily act as *dynamic contamination* for the *wind–windsea equilibrium*, resulting in large data scatter in measured $C_D$ values.

# 5 Nondimensional $CO_2$ transfer velocity as a function of the windsea Reynolds number

## 5.1 Nondimensional $CO_2$ transfer velocity

Both momentum and gas transfers across the sea surface proceed under the existence of wind and windsea. In the windsea boundary layer the laminar–turbulent transition is important, as discussed in Section 4.1. Momentum transfer has been expressed by eqn (2), using the nondimensional drag coefficient $C_D$, and we have seen that it is primarily controlled by the *windsea Reynolds number*, in situations free of conspicuous swell. Thus, $C_D$ is a measure that can indicate the delicate behavior of momentum transfer caused by the wind and the windsea state.

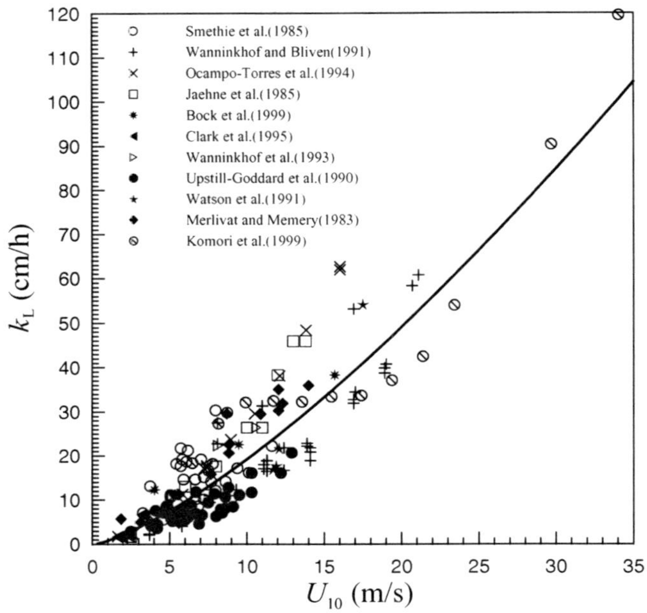


Figure 10: A synthesis of observed values of $CO_2$ transfer velocity $k_L$ from various sources, cited from Zhao *et al.* [27]. The distribution of data points indicates that $k_L$ already includes the factor for wind proportionality.

On the other hand, gas transfer has so far been treated by using a formulation based on the dimensional gas transfer velocity, $k_L$. Comparing eqn (5) for the definition of $k_L$ with eqn (2) for that of $C_D$, we infer that $k_L$ already includes the factor for wind proportionality, as is obviously seen from a synthesis of observed values of $k_L$ in Fig. 10, cited from Zhao *et al.* [27]. This is also inferred theoretically by Fairall *et al.* [13], in relation to their eqn (8).

We therefore propose the nondimensional transfer velocity for $CO_2$, defined by $k_L/U_{10}$, namely,

$$F = (k_L/U_{10})U_{10}S\Delta pCO_2. \tag{16}$$

In this way we remove the wind proportionality part, and may extract or amplify the delicate part of the behavior of $CO_2$ transfer, in order to closely examine the role of the *windsea Reynolds number* $R_B$, as in the case of $C_D$.

It should be noted that Zhao *et al.* [27] proposed a formula for $k_L$ as a function of $R_B$, as one step to the present paper, and Suzuki *et al.* [71] attempted a global mapping of the air–sea $CO_2$ exchange by using this formula.

## 5.2 Similarity and dissimilarity between momentum and gas transfers

Figure 11 shows a synthesis of observed values of the nondimensional transfer velocity $k_L/U_{10}$ expressed as a function of $R_B$, both for laboratory and

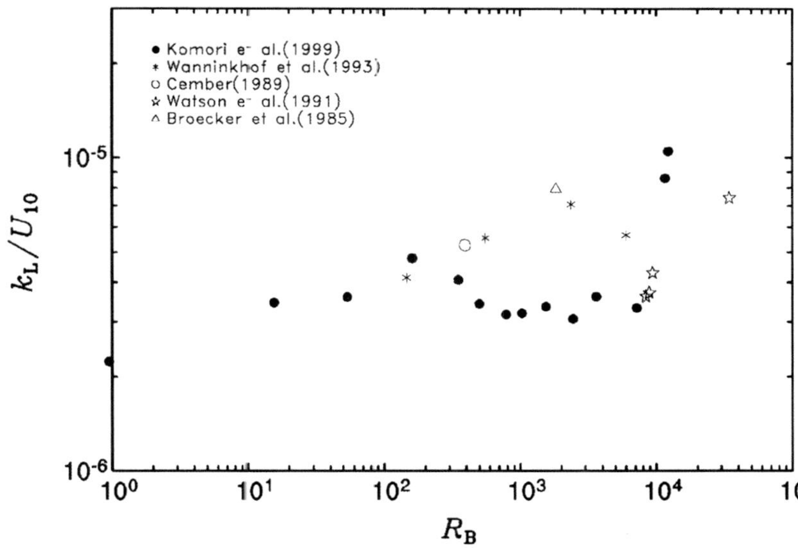


Figure 11: A synthesis of the observed values of the nondimensional transfer velocity $k_L/U_{10}$ expressed as a function of $R_B$, both for laboratory and field observations.

field observations. The number of data is small; nevertheless we can see important features in the behavior of $k_L/U_{10}$. As a matter of fact, data sets which included $k_L$, $u_*$, the wind speed at a certain level or $U_{10}$ or $C_D$, and windsea data are very rare. Though more data were included in Fig. 10, some of these, e.g., that of Jähne *et al.* [72], were shown by $U_{10}$ values estimated using an appropriate $C_D$ formula.

In Fig. 11, we first examine the laboratory data obtained by Komori *et al.* [26], shown by closed circles. These data are corrected, using a new calibrating curve, to compensate for errors due to the infrared $CO_2$ meter that have recently been found by the authors. We see that the $k_L/U_{10}$ values show a decrease between $R_B = 2 \times 10^2$ and $10^3$, corresponding to the $R_B$ range where $C_D$ shows a bulge. In the momentum transfer process, the pressure–force mechanism causing the form drag is important, in addition to the tangential stress causing the skin friction, as discussed at the end of Section 4.1 (Fig. 6). In the gas transfer process, however, the mechanism corresponding to the form drag does not exist. As the tangential stress is flat in this range of $R_B$, the gas transfer velocity should also be flat in the same range. This causes the decrease in $k_L/U_{10}$ as $U_{10}$ increases. This point is the conspicuous difference between the momentum and gas transfer processes.

In Fig. 11, we also see a second critical point beginning at $R_B = 10^3$ and shifting to the right (in $R_B$) by an order of magnitude, in the field data. This is just as in the case of $C_D$ (Fig. 7).

## 5.3 Similarity in laboratory and field data relating $C_D$ and $k_L$

Figure 12 shows $k_L/U_{10}$ plotted as function of $C_D$. Within the error bars of the data, the laboratory data and the field data seem to coincide with each other. The variation of $k_L$ with $C_D$ is shown in Fig. 13. In this case, the laboratory and the

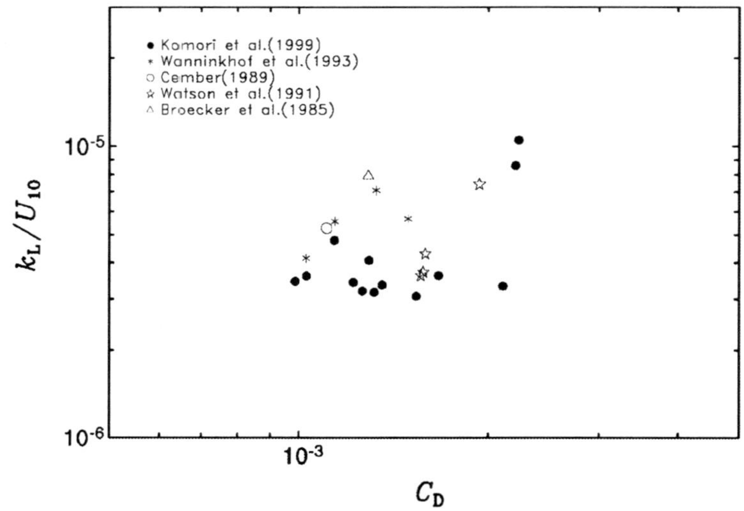


Figure 12: The variation of $k_L/U_{10}$ as a function of $C_D$. Data sources are same as in Fig. 11.

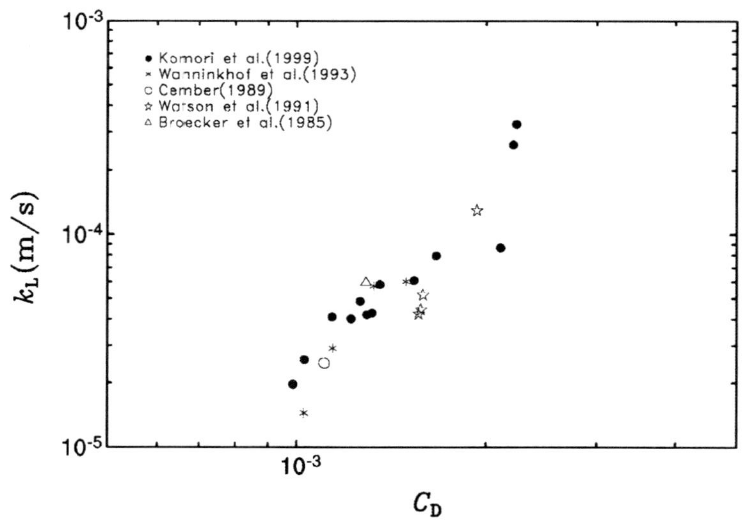


Figure 13: The same data as in Fig. 12, plotting $k_L$ as a function of $C_D$. Data sources are same as in Fig. 11.

field data show conspicuous agreement with each other. In this figure, the middle relatively flat part of the distribution of data points corresponds to the dissimilarity region in the momentum and gas transfers, between $R_B = 2 \times 10^2$ and $10^3$.

We interpret the difference between Figs. 12 and 13 as follows. In all data sets used in these figures, actual $U_{10}$ values were not necessarily measured directly. Rather, we used $U_{10}$ values that had been extrapolated from the wind speed measured at a certain level to 10-m values using the measured stress value with the logarithmic law. This procedure may cause part of the scatter shown in Fig. 11.

In summary, we suggest that there is a fundamental similarity between gas and momentum transfers. This includes our interpretation of the shift of critical points with regard to the *windsea Reynolds number* from laboratory to field data, occurring because of the fluid dynamical smoothing effect due to the intrinsic existence of swell components in the real sea, as discussed for Fig. 9.

## 6 Conclusions

We have deduced, for the first time, that conditions of windsea-dominant seas can be described by only two nondimensional parameters. One is the *wave age* for the representative windsea, or equivalently the nondimensional significant wave height or wave period, defined by eqn (14), under the existence of the 3/2-power law, eqn (6). The other is the *windsea Reynolds number* $R_B$, defined by eqn (15). The former describes the degree of windsea development relative to the local wind, whereas the latter is the fundamental controlling parameter for the *windsea boundary layer*, and consequently for the behavior of air–sea transfers.

Though the air–sea $CO_2$ transfer has traditionally been treated by use of transfer velocity $k_L$ as a function of the 10-m wind speed $U_{10}$, we propose the use of a nondimensional transfer velocity $k_L/U_{10}$, in correspondence with the drag coefficient $C_D$ for the momentum transfer. These nondimensional coefficients, $k_L/U_{10}$ and $C_D$, eliminate the primary wind-speed proportionality part in previous formulations and extract the detailed behavior of the processes of air–sea transfers. There is a fundamental similarity in the behavior of these two nondimensional coefficients, with some dissimilarity in a special range of $R_B$.

It has been shown that though $C_D$ generally increases with $R_B$, it has two critical points at $R_B = 2 \times 10^2$ and $10^3$ in the pure windsea case, with a flat bulge region between these. The first ($R_B = 2 \times 10^2$) corresponds to the transition from the region where tangential stress prevails, to the occurrence of airflow separation and reattachment to create the form drag. The second ($R_B = 10^3$) is a conspicuous critical point beyond which windsea breaking commences on the fully rough water surface. Namely, it is the transition to the full turbulence regime of the windsea boundary layer. On the other hand, $k_L/U_{10}$ decreases between $R_B = 2 \times 10^2$ and $10^3$. This is explained by the fact that in the gas transfer processes, the mechanism corresponding to the pressure–force in the momentum transfer does not exist.

It has also been deduced that field data under usual swell conditions show smaller transfer values or a remarkable shift of the second critical point to larger values in $R_B$. This is common for these two nondimensional transfer coefficients, $k_L/U_{10}$ and $C_D$. This corresponds to the observation that windsea energy becomes small in the presence of swell, especially in a direction different from the wind, which causes the sea surface to be dynamically smoother. In plots of $k_L/U_{10}$ as a function of $C_D$ and $k_L$ as a function of $C_D$, the distributions of points from windsea tank data and from field data coincide with each other well. These characteristics demonstrate the gross similarity in the momentum and the $CO_2$ transfers.

Thus, we propose further studies of the behavior of the nondimensional air–sea $CO_2$ transfer coefficient as a function of $R_B$. These studies should consider various kinds of swell and unsteady winds. Success in this endeavor would help push our understanding of air–sea fluxes into the mainstream of modern oceanography, in terms of the contemporary evolution of oceanic data acquisition systems with powerful numerical simulations and ocean-wave models.

## Acknowledgments

We appreciate the kind encouragement of Dr. William Perrie during this work and his valuable comments on the manuscript. The principal part of this chapter was presented as an invited paper, at a session of the Surface Ocean and Lower Atmosphere Study (SOLAS), convened by Professor Mitsuo Uematsu, within the IUGG General Assembly in Sapporo, July 2003. The authors would like to acknowledge the financial supports by a Grant-in-Aid for Scientific Research (S) (No.14102016) from the Japan Society for the Promotion of Science, and by the 21st Century Center of Excellence Program for Research and Education on Complex Functional Mechanical Systems.

## References

[1] Soloviev, A. & Schlüssel, P., Comments on "Air-sea gas transfer: Mechanisms and parameterization". *Journal of Physical Oceanography*, **28**, pp. 1643–1645, 1998.

[2] Csanady, G.T., *Air-Sea Interaction: Laws and Mechanisms*, Cambridge University Press: Cambridge, 2001.

[3] Toba, Y., Similarity laws of the wind wave and the coupling process of the air and water turbulent boundary layers. *Fluid Dynamics Research*, **2**, pp. 263–279, 1988.

[4] Toba, Y., Wind-forced strong wave interactions and quasi-local equilibrium between wind and windsea with the friction velocity proportionality (Chapter 1). *Nonlinear Ocean Waves (Advances in Fluid Mechanics, Vol. 17)*, ed. W. Perrie, Computation Mechanics Publications: Southampton, pp. 1–59, 1998.

[5] Toba, Y. & Koga, M., A parameter describing overall conditions of wave breaking, whitecapping, sea-spray production and wind stress. *Oceanic Whitecaps*, eds E.C. Monahan & G. MacNiocaill, D. Reidel Publishing Company: Dordrecht, pp. 37–47, 1986.

[6] Zhao, D. & Toba, Y., Dependence of whitecap coverage on wind and wind-wave properties, *Journal of Oceanography*, **57**, pp. 603–616, 2001.
[7] Iida, N., Toba, Y. & Chaen, M., A new expression for the production rate of sea water droplets on the sea surface, *Journal of Oceanography*, **48**, pp. 439–460, 1992.
[8] Yelland, M.J., Moat, B.I., Taylor, P.K., Pascal, R.W., Hatchings, J. & Cornell, V.C., Measurements of the open ocean drag coefficient for air flow disturbance by the ship. *Journal of Physical Oceanography*, **28**, pp. 1511–1526, 1998.
[9] Yelland, M.J. & Taylor, P.K., Wind stress measurements from the open ocean. *Journal of Physical Oceanography*, **26**, pp. 541–558, 1996.
[10] Jones, I.S.F. & Toba, Y. (eds). *Wind Stress over the Ocean*, Cambridge University Press: Cambridge, 2001.
[11] Suzuki, N., Ebuchi, N., Zhao, C., Watabe, I. & Sugimori, Y., Study of the relationship between non-dimensional roughness length and wave age, effect by wave directionality. *Proc. Indian Academy of Sciences (Earth and Planetary Sciences)*, **111(3)**, pp. 305–313, 2002.
[12] McGills, W.R., Edson, J.B., Hare, J.E. & Fairall, C.W., Direct covariance air-sea $CO_2$ fluxes. *Journal of Geophysical Research*, **106**, pp. 16729–16745, 2001.
[13] Fairall, C.W., Hare, J.B., Edson, J.B. & McGillis, W., Parameterization and micrometeorological measurement of air-sea gas transfer. *Boundary-Layer Meteorology*, **96**, pp. 63–105, 2000.
[14] Haubecker, H. & Jähne, B., In situ measurements of the air-sea gas transfer rate during the MBL/CoOP west coast experiment. *Air-Water Gas Transfer*, eds B. Jähne & E. Monahan, AEON: Hanau, pp. 553–569, 1995.
[15] Wanninkhof, R., Asher, W., Weppernig, R., Chen, H., Schlosser, P., Langdon, C. & Sambrotto, R., Gas transfer experiment on Georges Bank using two volatile deliberate tracers. *Journal of Geophysical Research*, **98**, pp. 20237–20248, 1993.
[16] Watson, A.J., Upstill-Goddard, R.C. & Liss, P.S., Air-sea gas exchange in rough and stormy seas measured by a dual-tracer technique. *Nature*, **349**, pp. 145–147, 1991.
[17] Cember, R., Bomb radiocarbon in the Red Sea. *Journal of Geophysical Research*, **94**, pp. 2111–2123, 1989.
[18] Broecker, W.S., Peng, T.H., Ostlund, G. & Stuiver, M. The distribution of bomb radiocarbon in the ocean. *Journal of Geophysical Research*, **90**, pp. 6953–6970, 1985.
[19] Komori, S. & Shimada, T., Gas transfer across a wind-driven air-water interface and the effects of sea water on $CO_2$ transfer. *Air-Water Gas Transfer*, eds B. Jähne & E. Monahan, AEON: Hanau, pp. 553–569, 1995.
[20] Komori, S. & Misumi, R., The effects of bubbles on mass transfer across the breaking air-water interface. *Gas Transfer at Water Surface*, eds E.S. Saltzman, M. Donelan, W. Drennan & R. Wanninkhof, AGU Monograph, pp. 285–290, 2002.

[21] Liss, P.S. & Merlivat, L., Air-sea gas exchange rates: introduction and synthesis. *The Role of Air-Sea Exchange in Geochemical Cycling*, ed. P. Buat-Menard, D. Reidel Hingham, MA, pp. 113–129, 1986.

[22] Wanninkhof, R., Relationship between wind speed and gas exchange over the ocean, *Journal of Geophysical Research*, **97**, pp. 7373–7382, 1992.

[23] Komori, S., Ueda, H., Ogino, F. & Mizushina, T., Turbulence structure and transport mechanism at the free surface in an open channel flow. *International Journal of Heat and Mass Transfer*, **25**, pp. 513–521, 1982.

[24] Komori, S., Murakami, Y. & Ueda, H., The relationship between surface-renewal and bursting motions in an open-channel flow. *Journal of Fluid Mechanics*, **203**, pp. 103–123, 1989.

[25] Komori, S., Nagaosa, R. & Murakami, Y., Turbulence structure and mass transfer across a sheared air-water interface in wind-driven turbulence. *Journal of Fluid Mechanics*, **249**, pp. 161–183, 1993.

[26] Komori, S., Shimada, T. & Misumi, R., Turbulence structure and mass transfer at a wind-driven air-water interface. *Wind-over-Wave Couplings*, eds S.G. Sajjadi, N.H. Thomas & J.C.R. Hunt, Oxford University Press: New York, pp. 273–285, 1999.

[27] Zhao, D., Toba, Y., Suzuki, Y. & Komori, S., Effect of wind waves on air-sea gas exchange: proposal of an overall $CO_2$ transfer velocity formula as a function of breaking-wave parameter. *Tellus*, **55B**, pp. 478–487, 2003.

[28] Toba, Y. Chapters 1 and 2. *Ocean-Atmosphere Interactions*, ed. Y. Toba, TERRAPUB/Kluwer: Tokyo/Dordrecht, pp. 1–62, 2003.

[29] Ebuchi, N., Toba, Y. & Kawamura, H., Statistical study on the local equilibrium between wind and wind waves by using data from ocean data buoy stations. *Journal of Oceanography*, **48**, pp. 77–92, 1992.

[30] Johnson, H.K., Hojstrup, H.J., Vested, H.J. & Larsen, S.E., On the dependence of sea surface roughness on wind waves. *Journal of Physical Oceanography*, **28**, pp. 1702–1716, 1998.

[31] Toba, Y., Local balance in the air-sea boundary processes—I. On the growth process of wind waves. *Journal of the Oceanographic Society of Japan*, **28**, pp. 109–120, 1972.

[32] Toba, Y., Stochastic form of the growth of wind waves in a single-parameter representation with physical implications. *Journal of Physical Oceanography*, **8**, pp. 494–507, 1978.

[33] Wilson, B.W., Numerical prediction of ocean waves in the North Atlantic for December 1959. *Dtsch. Hydrogr. Z.*, **18**, pp. 114–130, 1965.

[34] Mitsuyasu, H., Nakamura, R. & Komori, T., Observations of the wind and waves in Hakata Bay. *Rep. Res. Inst. Appl. Mech., Kyushu University*, **19**, pp. 37–74, 1971.

[35] Hasselmann, K., Barnett, T.P., Bouws, E., Carlson, H., Cartwright, D.E., Enke, K., Ewing, J.A., Gienapp, H., Hasselmann, D.E., Kruseman, P., Meerburg, A., Mueller, P., Olbers, D.J., Richter, K., Sell, W. & Walden, H., Measurements of wind-wave growth and swell decay during the Joint North

Sea Wave Project (JONSWAP). *Deut. Hydrogr. Z., Suppl. A*, **8(12)**, 95pp., 1973.

[36] Toba, Y., Iida, N., Kawamura, H., Ebuchi, N. & Jones, I.S.F., Wave dependence of sea-surface wind stress. *Journal of Physical Oceanography*, **20**, pp. 706–721, 1990.

[37] Toba, Y., Okada, K. & Jones, I.S.F., The response of wind-wave spectra to changing winds. Part I. Increasing winds. *Journal of Physical Oceanography*, **18**, pp. 1231–1240, 1988.

[38] Toba, Y., Jones, I.S.F., Ebuchi, N. & Kawamura, H., The role of gustiness in determining the aerodynamic roughness of the sea surface. *The Air-Sea Interface*, eds M.A. Donelan, W.H. Hui & W.J. Plant, University of Miami Press: Miami, pp. 407–412, 1996.

[39] Waseda, T., Toba, Y. & Tulin, M.P., Adjustment of wind waves to sudden changes of wind speed. *Journal of Oceanography*, **57**, pp. 519–533, 2001.

[40] Komatsu, K. & Masuda, A., A new scheme of nonlinear energy transfer among wind waves: RIAM Method—algorithm and performance. *Journal of Oceanography*, **52**, pp. 509–537, 1996.

[41] The WAMDI Group, The WAM model—a third generation ocean wave prediction model. *Journal of Physical Oceanography*, **18**, pp. 1775–1810, 1988.

[42] Suzuki, Y., Development and application of a global ocean wave prediction model including nonlinear interactions and dissipation, PhD Thesis, Graduate School of Science, University of Tokyo, 1995.

[43] Bailey, R.J., Jones, I.S.F. & Toba, Y., The steepness and shape of wind waves, *Journal of the Oceanographic Society of Japan*, **47**, pp. 249–264, 1991.

[44] Tokuda, M. & Toba, Y., Statistical characteristics of individual waves in laboratory wind waves. II. Self-consistent similarity regime. *Journal of the Oceanographic Society of Japan*, **38**, pp. 8–14, 1982.

[45] Toba, Y., Local balance in the air-sea boundary processes—III. On the spectrum of wind waves. *Journal of the Oceanographic Society of Japan*, **29**, pp. 209–220, 1973.

[46] Kawai, S., Okada, K. & Toba, Y., Field data support of the three-seconds power law and the $gu_*\sigma^{-4}$-spectral form for growing wind waves. *Journal of the Oceanographic Society of Japan*, **33**, pp. 137–150, 1977.

[47] Phillips, O.M., Spectral and statistical properties of the equilibrium range of wind generated gravity waves. *Journal of Fluid Mechanics*, **156**, pp. 505–531, 1985.

[48] Zakharov, V.E. & Filonenko, N.N., Energy spectrum for stochastic oscillation of the surface of a liquid. *Doklady Akademii Nauk SSSR*, **170**, pp. 1291–1295, 1966.

[49] Joseph, P.S., Kawai, S. & Toba, Y., Ocean wave prediction by a hybrid model—combination of single-parameterized wind waves with spectrally treated swells. *Tohoku Geophys. J. (Sci. Rep. Tohoku Univ. Ser. 5)*, **28**, pp. 27–45, 1981.

[50] Tulin, M.P., Breaking of ocean waves and downshifting. *Waves and Non-linear Processes in Hydrodynamics*, eds J. Grue, B. Gjevich & J.E. Weber, Kluwer: Dordrecht, pp. 170–190, 1996.

[51] Mitsuyasu, H. & Honda, T., The high frequency spectrum of wind-generated waves. *Journal of Physical Oceanography*, **30**, pp. 185–198, 1974.

[52] Toba, Y., Suzuki, Y. & Iida, N., Study on global distribution with seasonal variation of the whitecap coverage and sea-salt aerosol production on the sea surface. *The Wind Driven Air-Sea Interface*, ed. M.L. Banner, School of Mathematics, University of NSW, pp. 355–356, 1999.

[53] Monahan, E.C. & MacNiocaill, G. (eds). *Oceanic Whitecaps and Their Role in Air-Sea Exchange Processes*, D. Reidel: Dordrecht, 1986.

[54] Wu, J., Oceanic whitecaps and sea state. *Journal of Physical Oceanography*, **9**, pp. 1064–1068, 1979.

[55] Toba, Y. & Kawamura, H., Wind-wave coupled downward-bursting boundary layer (DBBL) beneath the sea surface. *Journal of Oceanography*, **52**, pp. 409–419, 1996.

[56] Toba, Y. & Kunichi, H., Breaking of wind waves and the sea surface wind stress. *Journal of the Oceanographic Society of Japan*, **26**, pp. 71–80, 1970.

[57] Banner, M. & Peirson, W., Tangential stress beneath wind-driven air-water interfaces. *Journal of Fluid Mechanics*, **364**, pp. 115–145, 1998.

[58] Hanson, J.L. & Phillips, O.M., Wind sea growth and dissipation in the open sea. *Journal of Physical Oceanography*, **29**, pp. 1633–1648, 1999.

[59] Kawai, S., Visualization of airflow separation over wind-wave crests under moderate wind. *Boundary-Layer Meteorology*, **21**, pp. 93–104, 1981.

[60] Kawai, S., Structure of air flow over wind wave crests. *Boundary-Layer Meteorology*, **23**, pp. 503–521, 1982.

[61] Kawamura, H. & Toba, Y., Ordered motion in the turbulent boundary layer over wind waves. *Journal of Fluid Mechanics*, **197**, pp. 105–138, 1988.

[62] Ebuchi, N., Kawamura, H. & Toba, Y., Fine structure of laboratory wind-wave surfaces studied using an optical method. *Boundary-Layer Meteorology*, **39**, pp. 133–151, 1987.

[63] Toba, Y., Drop production by bursting of air bubbles on the sea surface—III. Study by use of a wind flume. *Mem. College of Sci., University of Kyoto*, **A29**, pp. 313–344, 1961.

[64] Koga, M., Direct production of droplets from breaking wind-waves—its observation by a multi-colored overlapping exposure photographing technique. *Tellus*, **33**, pp. 552–563, 1981.

[65] Okuda, K., Internal flow structure of shot wind waves. Part I. On the internal vorticity structure. *Journal of the Oceanographic Society of Japan*, **38**, pp. 28–42, 1982.

[66] Shapiro, A.H., *Shape and Flow*, Educational Services Incorporated, originally published by Doubleday & Co., Inc.: New York, 1961.

[67] Koga, M., Bubble entrainment in breaking wind waves. *Tellus*, **34**, pp. 481–489, 1982.

[68] Hatori, M., Tokuda, M. & Toba, Y., Experimental study on strong interaction between regular waves and wind waves I. *Journal of the Oceanographic Society of Japan*, **37**, pp. 111–119, 1981.

[69] Toba, Y., Hatori, M., Imai, Y. & Tokuda, M., Experimental study of elementary processes in wind-waves using wind over regular waves. *Wave Dynamics and Radio Probing of the Ocean Surface*, eds O.M. Phillips & K. Hasselmann, Plenum: New York, pp. 117–127, 1986.

[70] Taylor, P.K. & Yelland, M.J., The dependence of sea surface roughness on the height and steepness of the waves. *Journal of Physical Oceanography*, **31**, pp. 572–590, 2001.

[71] Suzuki, Y., Komori, S., Zhao, D. & Toba, Y., Global mapping with seasonal variation of air-sea $CO_2$ exchange by using wind and wind-wave distribution. *Extended Abstracts, 6th International Carbon Dioxide Conference, Sendai*, Vol. II, pp. 756–759.

[72] Jähne, B., Wais, T., Memery, L., Caulliez, G., Merlivat, G.L., Münnich, K.O. & Coantic, M., He and Rn gas exchange experiments in the large wind-wave facility of IMST. *Journal of Geophysical Research*, **90**, pp. 11989–11997, 1985.

## Nomenclature

| | |
|---|---|
| $B$ | Proportionality factor for the 3/2-power law |
| $C_\mathrm{D}$ | Drag coefficient |
| $C_\mathrm{p}$ | Phase speed for peak frequency windsea |
| $C_\mathrm{s}$ | Phase speed for significant waves |
| $f_\mathrm{s}$ | Frequency of appearance of surface-renewal eddies |
| $g$ | Acceleration due to gravity |
| $H_\mathrm{s}$ | Significant wave height |
| $H^*$ | Nondimensional significant wave height |
| $k_\mathrm{L}$ | $CO_2$ transfer velocity |
| $k_\mathrm{s}$ | Significant wave number |
| $S$ | Solubility of $CO_2$ in water |
| $T_\mathrm{s}$ | Significant wave period |
| $T^*$ | Nondimensional significant wave period |
| $(\overline{u_a'^2})^{1/2}$ | Intensity of air turbulence |
| $(\overline{u_w'^2})^{1/2}$ | Intensity of water turbulence |
| $u_0$ | Stokes drift velocity |
| $u_\mathrm{s}$ | Averaged local wind drift |
| $u_*$ | Air friction velocity |
| $u_{*\mathrm{w}}$ | Water friction velocity |
| $W$ | Percentage of whitecap coverage |
| $z_0$ | Aerodynamic roughness parameter |
| $\alpha$ | Percentage of breaking crests of individual wind waves |
| $\alpha_\mathrm{s}$ | Coefficient of $gu_*\sigma^{-4}$-type one-dimensional windsea |

| | |
|---|---|
| $\beta$ | Wave age ($=C_s/u_*$ or $C_p/u_*$) |
| $\Delta p CO_2$ | Partial pressure difference of $CO_2$ between air and water |
| $\nu$ | Kinematic viscosity of air |
| $\sigma$ | Wave angular frequency |
| $\sigma_p$ | Spectral peak angular frequency |
| $\tau$ | Sea-surface wind stress |
| $\phi$ | Windsea spectral density |
| $\phi_p$ | Windsea spectral density at the peak frequency |

# CHAPTER 4

# Applications of synthetic aperture radar in marine meteorology

T.D. Sikora[1], G.S. Young[2], R.C. Beal[3], F.M. Monaldo[4] & P.W. Vachon[5]

*[1]Department of Earth Sciences, Millersville University, USA.*
*[2]Department of Meteorology, Pennsylvania State University, USA.*
*[3]SSARGASSO Associates, USA.*
*[4]Ocean Remote Sensing Group, Johns Hopkins University Applied Physics Laboratory, USA.*
*[5]Defense Research and Development Canada, Canada.*

## Abstract

This chapter reviews many of the marine meteorological capacities of synthetic aperture radar (SAR). We first examine the attributes of SAR image analysis in the study of air–sea interaction, providing examples of marine meteorological phenomena routinely imaged by SAR and discussions on how the scientific community can exploit this proven ability of SAR. Phenomena examined are organized by scale as follows: microscale cellular convection, microscale roll vortices, microscale gravity waves, mesoscale gravity waves, mesoscale convection, polar mesoscale cyclones, tropical cyclones, macroscale fronts, and extratropical cyclones. Next, we provide a review of recent advances in the transfer of SAR images to high-resolution (of the order of 100 m) near-surface wind speed images. Finally, we summarize the history of SAR as a meteorological tool and discuss its future.

The field of SAR meteorology is advancing at a steady pace. The material presented in this chapter represents the state of the art as of early 2004.

## 1 Introduction

For more than two decades, it has been known that imaging microwave radar, such as synthetic aperture radar (SAR), can be employed as a marine meteorological tool (e.g., Beal *et al.* [1]). This chapter outlines some of the realized and potential meteorological capacities of SAR. We will first examine the attributes of SAR image analysis in the study of air–sea interaction, providing examples of marine meteorological phenomena routinely imaged by SAR and discussions on how the scientific community can exploit this proven ability of SAR. This will be followed by a review of recent advancements in the transfer of SAR images to high-resolution (of the order of 100 m) near-surface wind speed images. The potential uses of such a wind speed data set to those interested in marine meteorology are innumerable.

Before proceeding, we will present a brief review of the horizontal scales of atmospheric processes and turbulent transfer. An understanding of the horizontal scales of atmospheric processes is necessary to place any one meteorological phenomenon in the proper context with respect to others in this chapter. Turbulent transfer lies at the heart of SAR's ability as a meteorological instrument.

### 1.1 Horizontal scales of atmospheric processes

The following horizontal scale definitions are taken from Orlanski [2] and Stull [3]. See figure 1 from Orlanski [2] for a pictorial description of the following discussion. We begin with the macroscale, which is divided into two subranges: Macro $\alpha$ or planetary scale motions have horizontal spatial scales greater than 20 000 km. Examples of macro $\alpha$ phenomena are jet streams that circumnavigate a hemisphere. Proceeding towards smaller processes, the next scale encountered is macro $\beta$, the synoptic scale, which lies between 20 000 and 2000 km. The extratropical cyclone is an example of a macro $\beta$ circulation.

Next, we encounter mesoscale meteorological phenomena, which are divided into three groups: with spatial scales between 200 and 2000 km, meso $\alpha$ circulations include phenomena like hurricanes, polar mesoscale cyclones, and mesoscale fronts. Meso $\beta$ features have spatial scales between 200 and 20 km. Mesoscale convective complexes are often meso $\beta$ scale. Rounding out the mesoscale are meso $\gamma$ phenomena, which have spatial scales between 20 and 2 km and include thunderstorms and some of the larger atmospheric gravity waves.

Finally, we reach the microscale, beyond which is molecular dissipation. There are four microscale groups: micro $\alpha$ phenomena, with scales from 2 to 0.2 km, include boundary layer cumulus clouds, tornadoes, and yet more atmospheric gravity waves. Micro $\beta$ phenomena have spatial scales from 0.2 to 0.02 km. Dust devils and thermals are examples of such. The micro $\gamma$ scale lies between 0.02 and 0.002 km. Surface layer plumes are micro $\gamma$ scale. Lastly, the micro $\delta$ scale is encountered whose phenomena range between 0.002 and 0.0002 km in spatial scale. Small-scale mechanical turbulence is an example of a micro $\delta$ phenomenon.

### 1.2 Turbulent transfer and SAR

Turbulence, the irregular chaotic nature of many flows, is of particular importance to the utility of SAR as a meteorological instrument. Turbulence is quite ubiquitous in the portion of the atmosphere that is adjacent to the earth's surface. It is fueled by microscale gradients in momentum, temperature, and moisture, and its role is to destroy those very same gradients that give it life. This destruction occurs via turbulent transfer processes (fluxes), without which the transports of energy, moisture, and momentum between the earth's surface and the atmosphere would be left to molecular diffusion, having scales 3–6 orders of magnitude smaller than turbulent diffusion. Hence, turbulent fluxes fuel much of the larger-scale meteorology (e.g., Stull [3]).

The visible effects of turbulent fluxes are many. For example, cat's paws crossing a body of water result from turbulent momentum flux at the air–water interface. Herein lies the connection between turbulence and SAR's role as a meteorological tool: In the early part of the last century, Sir William Bragg demonstrated that the periodic structure of a crystal lattice produces constructive interference in reflected radiation resulting in an increase in the reflected energy when the crystal spacing matches the wavelength of the incident radiation. In 1960, Wright [4] applied the same principle to the reflection of energy from the ocean surface. When a radar illuminates the ocean surface at moderate incident angles (20–60 degrees), the dominant portion of the reflected power is produced by ocean surface roughness on the scale of the radar wavelength projected on to the ocean surface, the "Bragg" wavelength. Typical microwave radars operate at wavelengths on the decameter and centimeter scales and so the wind generated roughness (via the turbulent momentum flux) on these scales is responsible for the ocean surface radar signature. As the near-surface wind speed increases so does the surface roughness and consequently the backscattered power increases.

The short waves generated by the wind dominantly travel in the along wind direction. For this reason, the reflected electromagnetic energy is a maximum when the local wind is pointing into the radar look direction. There is a similar, though somewhat smaller local maximum in the reflected power when the wind is blowing away from the radar. The minimum in the reflected power occurs when the radar look direction is perpendicular to the wind direction.

Thus, SAR senses the forcing that atmospheric phenomena exert on the centimeter-scale wave spectrum. At the same time, the intervening atmosphere is mainly transparent to SAR although precipitation can at times affect the radar signal (Melsheimer *et al.* [5, 6]).

The typical resolution of spaceborne SAR is of the order of 10–100 m with a swath width of the order of 100–1000 km (Mourad [7]). Thus, spaceborne SAR is capable of providing a detailed view of sea-surface stress-induced roughness patterns (the footprints) of macroscale, mesoscale, and microscale meteorological phenomena.

We point out that all figures labeled as "SAR images" contained herein have been scaled for presentation. In particular, a systematic range-dependent trend caused by the antenna beam pattern has been removed from the images. Thus, the reader should not conclude that the gray scale of any SAR image presented below represents backscattered normalized radar cross section (NRCS). Instead, look upon the data as generic intensity.

# 2 SAR image analysis in the study of marine meteorological phenomena

Here, we provide examples of marine meteorological phenomena routinely imaged by SAR and information on how the scientific community can exploit this proven ability of SAR. The following discussion is organized by horizontal scale.

## 2.1 Microscale phenomena

We now outline some of the more common microscale marine atmospheric boundary layer (MABL) quasi-two dimensional signatures seen in SAR images, as outlined in Sikora and Young [8]. The ability of SAR to sense the footprint of a given microscale phenomenon provides a means by which one can infer the corresponding dynamic and thermodynamic environment associated with that phenomenon's existence (e.g., statically unstable versus statically stable; baroclinic versus barotropic), and thus may be of interest to those conducting MABL research, such as large eddy simulation studies and phenomenon climatologies, and to operational marine weather forecasters.

We will concentrate on the typical range of near-surface mean wind directions with respect to signature orientation. This is because, as discussed below in Section 3, recently there has been much effort put into the attempt to extract subkilometer scale near-surface wind speed estimates from SAR images using scatterometer-like transfer functions (discussed in Section 3). As is pointed out by Monaldo *et al.* [9], the near-surface mean wind direction is a required input for this transfer, and many researchers have based their determination of the near-surface mean wind direction on the orientation of the SAR signatures of quasi-two dimensional MABL phenomena.

However, as will be shown below, there is often a wide range of quasi-two dimensional MABL phenomena depicted in SAR images. As such, there can exist large differences in signature orientation with respect to the near-surface mean wind direction. Because it can be difficult to discern one such phenomenon from another in a SAR image, simple analysis of the orientation of the quasi-two dimensional SAR signatures will at times fail to yield the correct near-surface mean wind direction. Thus, we will also provide the reader with some empirically derived tips on how to discern one feature from another.

### 2.1.1 Convective cells

Under relatively light wind conditions and statically unstable stratification, a field of cellular convective updrafts and downdrafts is apt to form. The convective downdrafts tend to mix down relatively high-momentum air from near the top of the convective MABL towards the surface, leading to increased surface layer wind shear and increased turbulent momentum transfer to the sea surface. Over water, the increased momentum transfer results in increased centimeter-scale roughness and increased SAR intensity. The convective updrafts lead to decreased surface layer shear and decreased momentum transfer to the sea surface. This results in decreased centimeter-scale sea surface roughness and decreased SAR intensity. The resulting SAR intensity pattern beneath the field of cellular convection takes on a mottled appearance (e.g., Sikora *et al.* [10]; Zecchetto *et al.* [11]; Babin *et al.* [12]); and Fig. 1).

Under extremely light wind conditions throughout the depth of the convective MABL, the shape of a mottle element will be more or less circular because the air emanating from the downdraft at the surface spreads out radially in all directions due to continuity. In the presence of vertical wind shear throughout the depth of the convective MABL, an individual mottle element will tend to be elongated along the direction of the shear vector between the anemometer level (of the order of 10 m above sea level) and the top of the MABL. In the case of barotropic or weakly baroclinic convective MABLs, one can expect minimal directional shear and, thus, the mottles tend to be elongated along, or to within a few degrees clockwise of, the near-surface mean wind direction (e.g., Zecchetto *et al.* [11]). In the case of moderately to strongly baroclinic MABLs, there can be a large amount of directional shear across the convective MABL and, thus, the orientation of the mottles can be quite different from the near-surface mean wind direction. During moderate cold air advection events, the mottles will be oriented along, or to within 10–20 degrees

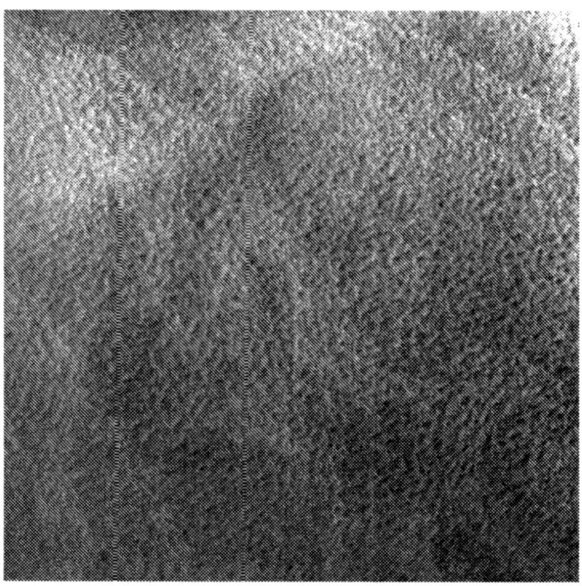

Figure 1: Radarsat-1 SAR image depicting the mottled signature of kilometer-scale cellular convection throughout. The 300 m pixel image is approximately 270 km × 270 km. The image was acquired at C-band, horizontal polarization, off the northeast coast of the United States at 2242 UTC on March 6, 1997. The top of the image is directed towards 348°T [Provided by JHUAPL, © Canadian Space Agency (CSA)].

counterclockwise of, the near-surface mean wind direction. During moderate warm air advection events, the mottles will be oriented clockwise of the near-surface mean wind direction, potentially by several tens of degrees.

### 2.1.2 Buoyancy-driven/shear-organized roll vortices

As their name implies, buoyancy-driven/shear-organized rolls are helical circulations that form via thermodynamic instability in an environment with sufficient vertical wind shear. For a given amount of MABL buoyancy, as the magnitude of the wind shear increases, a field of randomly organized elongated mottle elements evolves into a field of linearly organized elongated mottle elements. Ascending and descending regions of the circulation lead to the corresponding increased and decreased sea surface roughness in the same manner as was described for cellular convection (e.g., Müller *et al.* [13]). The resulting SAR intensity pattern takes on an appearance of alternating dark and bright mottled lines (e.g., Alpers and Brümmer [14]; Babin *et al.* [12]; and Fig. 2). The orientation of the surface footprint of this type of roll, and thus its SAR signature, is forced in the same manner as was discussed previously for cellular convection (Weckwerth *et al.* [15, 16]).

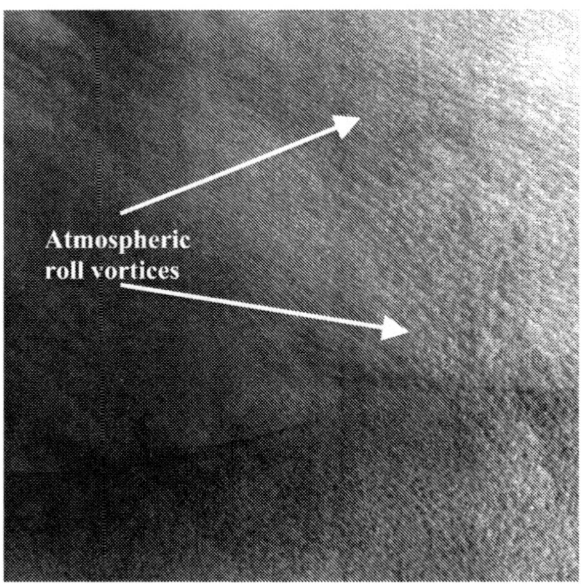


Figure 2: Radarsat-1 SAR image depicting the signature of roll vortices. The 300 m pixel image is approximately 270 km × 270 km. The image was acquired at C-band, horizontal polarization, off the northeast coast of the United States at 2242 UTC on March 6, 1997. The top of the image is directed towards 348°T (Provided courtesy of JHUAPL, © CSA).

### 2.1.3 Inflection-point-induced rolls

Inflection-point induced rolls form when inflection points exist in the vertical profile of the horizontal wind at times of neutral static stability (e.g., Brown [17]; Stensrud and Shirer [18]). Thus, inflection point rolls are purely shear driven, gaining their energy not from buoyancy but from the kinetic energy of the mean environment. As such, they tend to be aligned perpendicular to the shear vector at the inflection point (e.g., Stensrud and Shirer [18]). Ascending and descending regions of the circulation lead to the corresponding increased and decreased sea surface roughness in the same manner as was described for cellular convection. However, the modeling analysis provided in Müller *et al.* [13] suggests that the SAR signature of inflection-point-induced rolls should lack the mottled string-of-pearl appearance typical of buoyancy-driven/shear-organized rolls.

For unidirectional flow, the SAR signature of the rolls lies perpendicular to the near-surface mean wind direction. In a typical Ekman environment, the orientation of the signature of the rolls would be about 45 degrees clockwise of the near-surface mean wind direction (Stensrud and Shirer [18]). Warm advection (e.g., figure 3 from Alpers and Brümmer [14]) and cold advection can impact this relationship by tens of degrees.

### 2.1.4 Shear-driven gravity waves

Atmospheric gravity waves form in a stably stratified atmosphere when the vertical shear becomes sufficient to provide energy at a rate faster than it can be dissipated. When such waves form on a surface-based or low-altitude elevated inversion, they can result in perturbations of the near-surface wind speed. The resulting surface stress variations produce a banded pattern on SAR images, with the greatest roughness and corresponding SAR intensity under the wave troughs. This pattern, like the waves responsible, is aligned perpendicular to the shear across the inversion (e.g., Vachon *et al.* [19]). The SAR signature of gravity waves is expected to be even less variable along any one linear feature than that associated with inflection-point-induced rolls.

Such atmospheric internal gravity wave signatures are commonly observed on SAR images when mesoscale or synoptic scale frontal inversions approach the surface. Thus, they tend to occur near, but on the cool side of, the surface frontal position and tend to be more well defined, the stronger the front. For slowly moving fronts, the flow is nearly geostrophic so the vertical wind shear is roughly parallel to the front. Thus, the resulting atmospheric internal gravity waves are roughly perpendicular to the front. In contrast, for fast moving fronts, the flow is highly ageostrophic. Thus, both the near-surface mean wind and the vertical wind shear are quasi-perpendicular to the front. The resulting gravity waves parallel the front (e.g., Fig. 3). In either situation, the wave signatures fade out with distance from the surface front because of the increasing elevation of the frontal inversion. Thus, smooth, uniform bands of enhanced SAR intensity aligned perpendicular or parallel to a front and extending from near the front to of the order of 100 km to the cool side of the front should be suspected of being the result of atmospheric internal gravity waves, not roll vortices.

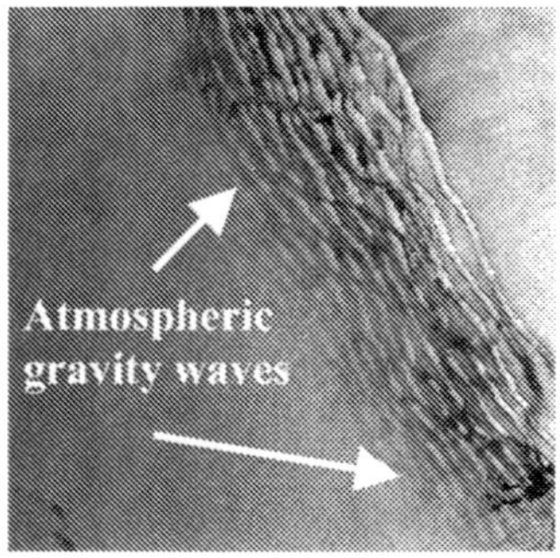


Figure 3: ERS-1 SAR image depicting the signature of atmospheric gravity waves associated with highly ageostrophic flow near a front. The 180 m pixel image is approximately 90 km × 90 km. The image was acquired at C-band, vertical polarization, over the Caspian Sea at 0723 UTC on May 12, 1996. The top of the image is directed towards 012°T [Provided courtesy of Werner Alpers and the European Space Agency (ESA), © ESA].

Given that near-surface mean winds on the cold side of strong (i.e., fast moving) cold fronts generally intersect the frontal surface at an angle of nearly 90 degrees, the SAR signatures of the corresponding shear-driven gravity waves are likely to be aligned more or less perpendicular to the near-surface mean wind direction. The same quasi-perpendicular relationship between the near-surface mean wind direction and the shear-driven gravity wave signature alignment also holds for slow moving fronts (warm, cold, or stationary) because the near-surface mean wind and the vertical shear are both more or less parallel to the front.

## 2.2 Mesoscale phenomena

Here, we will examine SAR's ability to sense the sea surface footprints of topographically driven gravity waves, mesoscale convection, polar mesoscale cyclones, and hurricanes. Where applicable, we provide information on the range of expected near-surface mean wind directions associated with each phenomenon.

As with microscale phenomena, the existence of these mesoscale SAR signatures can be used to infer the corresponding dynamic and thermodynamic environment, and the near-surface wind direction. Moreover, SAR provides unprecedented detail of the microstructure of each phenomenon. Thus, those interested in simulating and forecasting mesoscale atmospheric environments should find SAR a useful verification and analysis tool.

### 2.2.1 Topographically driven gravity waves

The SAR signatures of atmospheric gravity waves are also common when stably stratified air flows over the terrain (e.g., Winstead *et al.* [20]). The signatures appear to the lee of the terrain with ridges producing waves aligned parallel to their crests

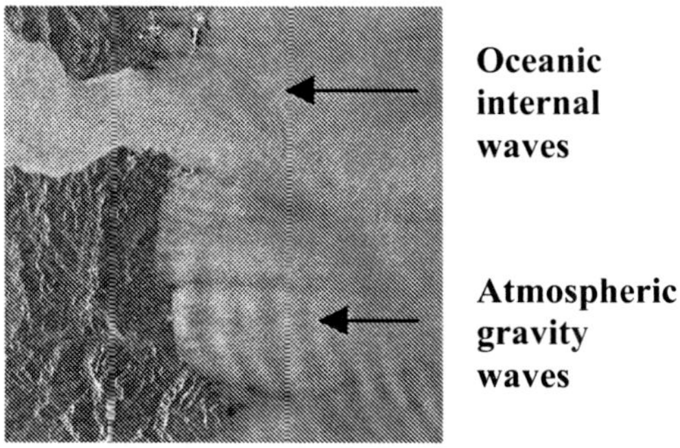


Figure 4: ERS-1 SAR image depicting the signature of atmospheric gravity waves forced by topography. The semi-circular wave packet seen near the top center portion of the image is the SAR signature of oceanic internal waves (e.g., Beal *et al.* [1]). The 180 m pixel image is approximately 90 km × 90 km. The image was acquired at C-band, vertical polarization, over the western Mediterranean Sea at 2239 UTC on September 3, 1993. The top of the image is directed towards 348°T (Provided courtesy of Werner Alpers and ESA, © ESA).

(e.g., lower right portion of Fig. 4) and isolated peaks producing v-shaped chevrons pointing upwind (e.g., upper right portion of Fig. 5). Each high intensity area in the SAR signature corresponds to a band of enhanced near-surface wind speed where the wave trough touches the sea surface. The low intensity areas correspond to the wave crests, wherein the strongest winds lift away from the surface.

The existence of smooth ridge-parallel SAR wave signatures implies conditions favorable for the formation of mountain lee waves, including near-surface mean winds oriented within 45 degrees of the perpendicular to the ridge. The orientation of the chevron wave pattern from an isolated peak indicates the direction of the winds near the height of the mountain. In cold advection, the near-surface mean winds could be tens of degrees counterclockwise from this mountaintop wind direction while in warm advection they could be tens of degrees clockwise.

### 2.2.2 Mesoscale convective cells

Cellular convection also occurs on the meso $\beta$ and meso $\gamma$ scales. Unlike microscale convective cells that can be either clear or cloudy, mesoscale convective cells appear to be associated exclusively with cumuloform clouds (i.e., cumulocongestus, cumulonimbus, and stratocumulus). As with microscale convection, however, the sea surface stress associated with mesoscale convective cells (Young *et al.* [21]), and thus their SAR signatures (e.g. Atlas [22]; Babin *et al.* [12]) result from downdraft modification of the surface wind field. The primary mechanism for this modification is the spreading of the downdraft air along the surface resulting in a quasi-circular signature (e.g., Fig. 6). The sharp edge of this signature corresponds to the edge of this outflow (i.e., the gust front).

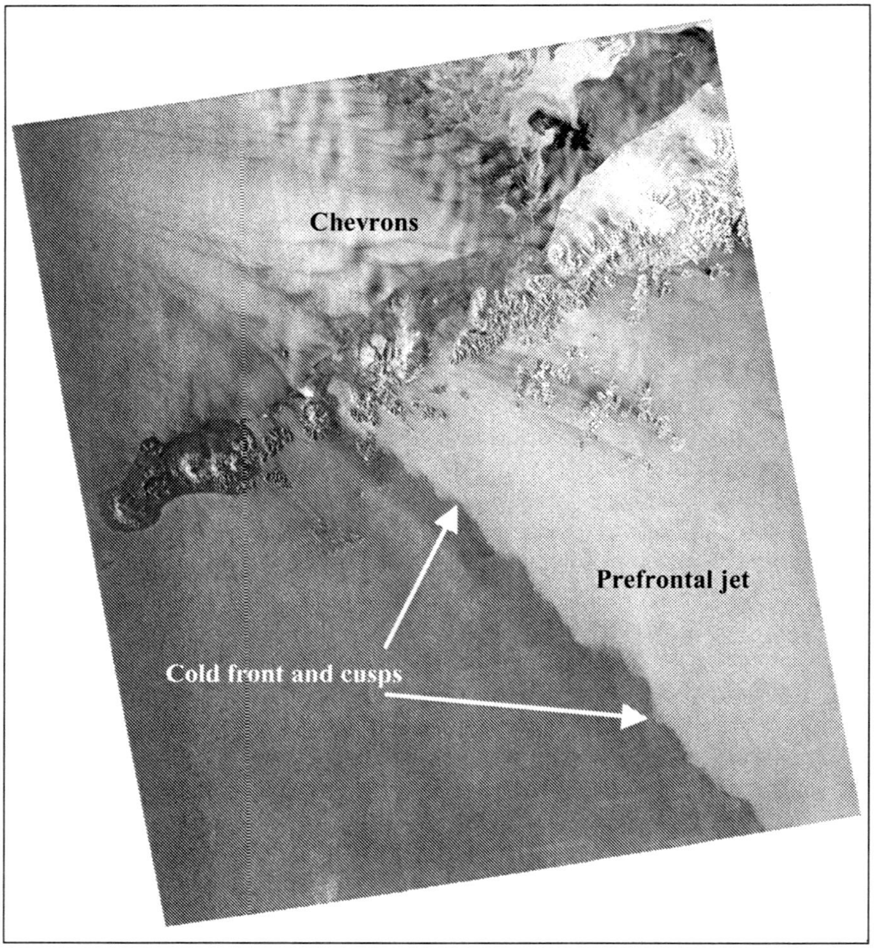


Figure 5: Radarsat-1 SAR image depicting the signature of a front, prefrontal jet, and frontal cusps. In addition, numerous chevron signatures can be seen. The 300 m pixel image is approximately 500 km × 415 km. The image was acquired at C-band, horizontal polarization, over the Alaska Peninsula at 0429 UTC on February 5, 2000. The top of the figure is directed towards 000°T (Provided courtesy of JHUAPL, © CSA).

Unlike microscale convection, which forms a densely packed array of signatures, mesoscale cellular convection typically produces more widely scattered signatures (e.g., Fig. 1 versus Fig. 6). This difference in downdraft coverage corresponds with that observed between nonprecipitating and precipitating convection (Gaynor and Mandics [23]), implying that many (if not all) mesoscale cellular convection signatures on SAR are the result of precipitation-driven downdrafts. The existence of scattered mesoscale convective signatures thus implies the existence of moist

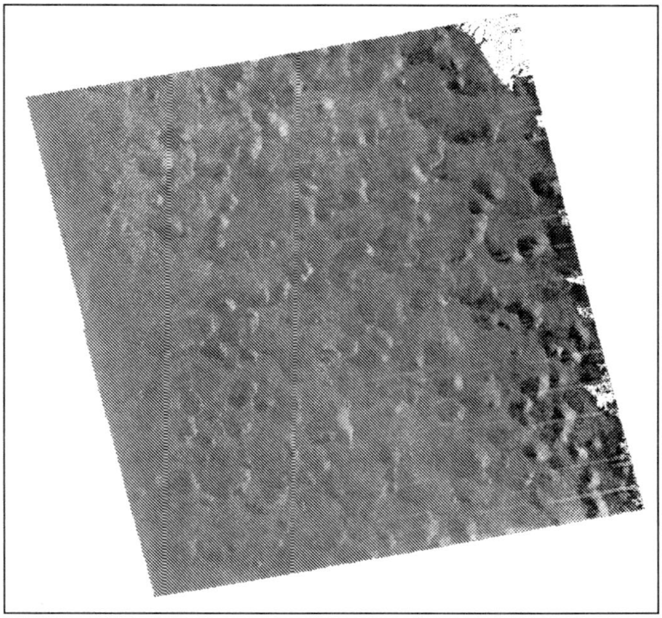

Figure 6: Radarsat-1 SAR image depicting the signature of mesoscale cellular convection. The 450 m pixel image is approximately 450 km × 450 km. The image was acquired at C-band, horizontal polarization, over the Gulf of Alaska at 0252 UTC on April 5, 2001. The top of the figure is directed towards 000°T (Provided courtesy of JHUAPL, © CSA).

precipitating convection and convective available potential energy in at least the lower troposphere. Further study may reveal a relationship between the size and spacing of the signatures and the depth of the unstable layer.

Because convective downdrafts modify the surface wind field via both surface divergence of the downdraft and the vertical transport of horizontal winds, the orientation of the resulting signatures reflects both the downdraft intensity and the winds aloft. Stronger downdrafts will result in a greater difference in the near-surface wind speed from the downwind to the upwind edge of the signature (e.g., the strong signatures in the center of Fig. 6 as contrasted with the weaker ones in the upper left and lower left corners). Transfer of winds from aloft causes the axis of this dipole to depart from that of the surface wind, thus providing an indication of the wind direction aloft. However, the inversion of this relationship is complicated by the interaction of the look-angle dependence of SAR backscatter with the diffluence of the surface outflow. The resulting orientation error caused by assuming uniform directional flow could be tens of degrees if the mean near-surface wind speeds were small relative to the divergent component of the outflow velocity.

### 2.2.3 Polar mesoscale cyclones

Polar mesoscale cyclone (PMC) is the generic term for all meso $\alpha$ and meso $\beta$ scale cyclonic vortices poleward of the polar front (Heinemann and Claud [24]). These intense cyclones form under a wide range of conditions (e.g., baroclinic instability, air–sea interaction instability, conditional instability of the second kind, or a combination of mechanisms), are rather short-lived, and produce strong winds, heavy precipitation, and large air–sea fluxes of sensible and latent heat (e.g., Bresch *et al.* [25]; Nielsen [26]; Miner *et al.* [27]). Thus, the proper analysis and forecasting of PMCs is of particular importance to polar marine commerce such as the Alaskan fishery industry.

Recently, SAR has been shown to be an effective means of providing high-resolution remote sensing data of PMCs. For example, Chunchuzov *et al.* [28] present a SAR-based study of PMCs in the Labrador Sea. Sikora *et al.* [29] and Friedman *et al.* [30] provide complementary studies of PMCs found in the Bering Sea. Here, we will summarize one of the PMC cases discussed by Sikora *et al.* [29].

Figure 7 is a Radarsat-1 SAR image of the sea surface footprint of a PMC over the Bering Sea. Dramatic SAR intensity boundaries spiral inward cyclonically towards the center of the PMC. Bresch *et al.* [25], Bond and Shapiro [31], and Douglas *et al.* [32], show cloud and wind features analogous to these intensity boundaries in their non-SAR studies. Their features are associated with confluence and/or frontal zones.

An isolated area of low SAR intensity is found at the center of the PMC. Bresch *et al.* [25] and Miner *et al.* [27] document isolated areas of low near-surface wind speed at the center of PMCs in their non-SAR studies and have attributed them to warm cores and thus increased surface layer stability and decreased air–sea interaction. PMC warm cores can result from warm air seclusion and/or adiabatic compression (Montgomery and Farrell [33]).

Mesoscale and microscale structures abound in and around the PMC shown in Figure 7. On the mesoscale, 20 km wave-like features (cusps), reminiscent of lobe and cleft instability (Lee and Wilhelmson [34]) exist along one of the spiral arms noted above. Chunchuzov *et al.* [28] show similar features along regions of large wind gradients associated with their PMCs. We will demonstrate in Section 2.3 that such features are common along the SAR signature of synoptic scale cold fronts.

As for microscale structure, notice that along the southern edge of the PMC's center (see inset within Fig. 7), 2 km alternating lines of high and low SAR intensity are apparent. These features are reminiscent of the SAR signature of MABL gravity waves (e.g., Vachon *et al.* [19]). The mottled SAR intensity pattern associated with MABL cellular convection (Sikora *et al.* [10]; Zecchetto *et al.* [11]) is apparent in the lower left hand corner of Fig. 7. Finally, the SAR signature of roll vortices can be seen within the top center of Fig. 7. We refer the reader to Section 2.1 for further interpretation of these SAR signatures as well as the range of near-surface wind directions associated with them.

### 2.2.4 Tropical cyclones

Tropical cyclones are potentially destructive storms that occur over some of the warmest of the Earth's oceans; they are locally known as typhoons in the western

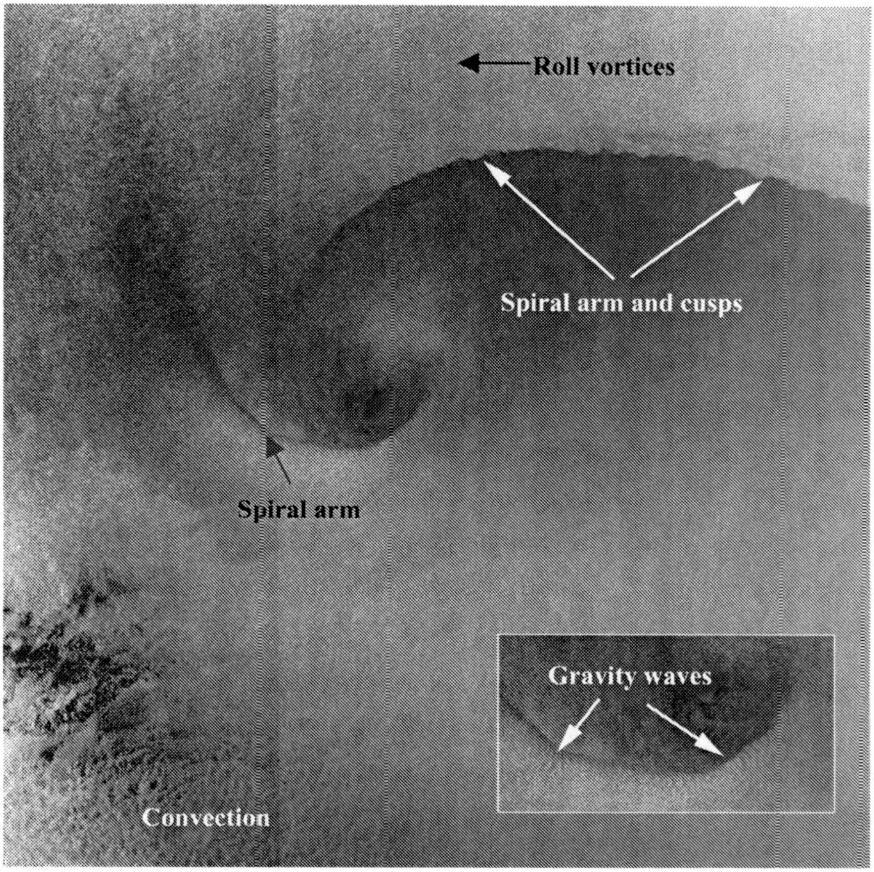


Figure 7: Radarsat-1 SAR image depicting the signature of a PMC. The 250 m pixel image is approximately 420 km × 420 km. The image was acquired at C-band, horizontal polarization, over the Bering Sea at 0602 UTC on February 5, 1998. The top of the image is directed towards 348°T (© CSA).

Pacific Ocean and as hurricanes in the Atlantic Ocean, Caribbean, and eastern Pacific. Tropical cyclones form via air–sea interaction with this warm water, with the resulting sensible and latent heat transferred from the atmospheric boundary layer to the troposphere via deep convection. Air–sea interaction is not the only aspect involved in tropical cyclone dynamics however, as they must form far enough away from the equator for the force of the Earth's rotation (Coriolis) to convert the thermally direct circulation of deep convection into a balanced vortex.

Both the convection and the resulting vortex yield distinctive signatures in SAR images (e.g., Fig. 8, Hurricane Erin in 2001) because of their impact on the air–sea flux of momentum. The vortex appears as a quasi-circular annulus of enhanced SAR intensity (high winds) surrounding a low-wind center (the hurricane eye

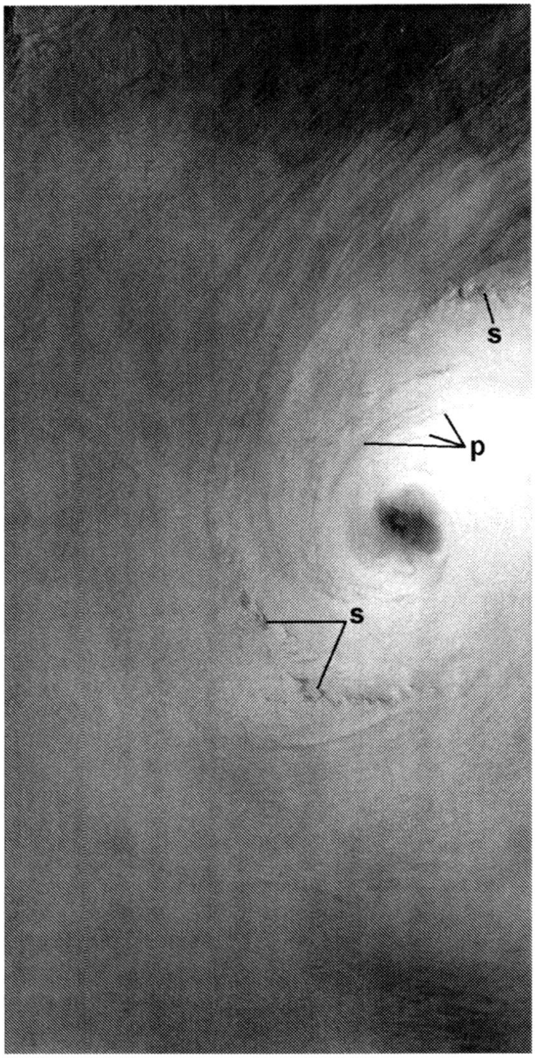


Figure 8: Radarsat-1 SAR image of Hurricane Erin. The 500 m pixel image is approximately 900 km × 450 km. The image was acquired at C-band, horizontal polarization, off the east coast of the United States at 2218 UTC on September 11, 2001. The top of the figure is directed towards 348°T. The eye is clearly evident; precipitation bands (p) and squall lines (s) are indicated (© CSA).

or its precursor in weaker cyclones). The convection results in both low- and high-SAR intensity features superimposed on the vortex signature. Because the vortex's deformation field stretches convective clusters, the convective signatures appear as a series of discrete updraft and downdraft signatures along a spiral band

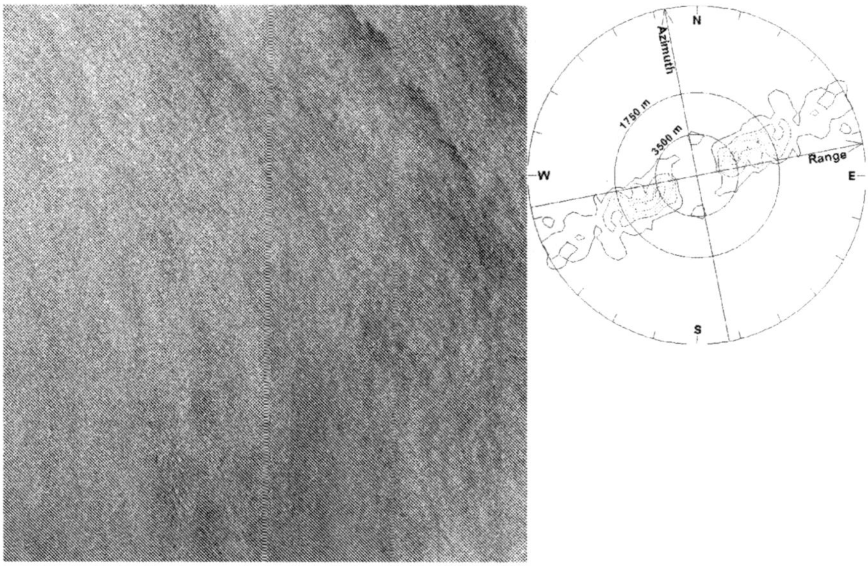


Figure 9: Radarsat-1 SAR image of Hurricane Floyd's boundary layer rolls (left). The 400 m pixel image is approximately 375 km × 350 km. The image was acquired at C-band, horizontal polarization, off the east coast of the United States at 2249 on September 14, 1999. The top of the image is directed towards 348°T (© CSA). The image spectrum (right) illustrates the scale of the streaks in the image (with 180° directional ambiguity), in this case about 3.5 km (adapted from Katsaros *et al.* [36]).

[labeled as squall lines (s) in Fig. 8]. Precipitation features usually appear darker (i.e. as having low SAR intensity), perhaps in part due to absorption by the rain, but also due to destruction of the wind-driven patterns of surface roughness by impacting raindrops. Thus, precipitation adds discrete low-SAR intensity elements along convective bands and continuous arcs of reduced backscatter along stratiform rainbands. The origin of these features is understood due to comparison of SAR images with contemporaneous coastal weather radar images of rainfall (Katsaros *et al.* [35]). SAR however provides much greater insight into processes at work at the sea surface than do conventional visible and infrared satellite images which show only the upper tropospheric cloud top. Thus, high-resolution SAR images of the impact of tropical cyclones on the ocean surface roughness distribution have provided new insight into tropical cyclone structure and dynamics (Katsaros *et al.* [35, 36]).

One potentially important discovery is the presence of the SAR signature of longitudinal roll vortices within many tropical cyclones. Evidence of these roll vortices are illustrated in an image of Hurricane Floyd (Fig. 9). These data are from a region between rainbands, roughly 500 km away from the eye. Because roll vortices affect the transfer of heat and moisture from the sea surface up through

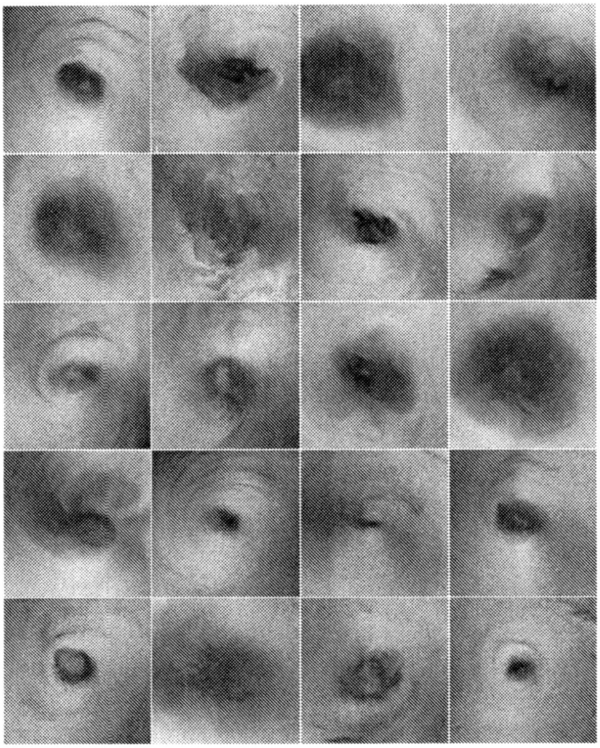

Figure 10: Representative Radarsat-1 SAR images of hurricane eyes, arranged as follows:

| Danielle<br>August 31, 1998 | Dennis<br>August 27, 1999 | Dennis<br>August 29, 1999 | Dennis<br>August 31, 1999 |
|---|---|---|---|
| Floyd<br>September 15, 1999 | Alberto<br>August 17, 2000 | Florence<br>September 13, 2000 | Dalila<br>July 26, 2001 |
| Flossie<br>August 29, 2001 | Flossie<br>September 1, 2001 | Erin<br>September 11, 2001 | Erin<br>September 13, 2001 |
| Felix<br>September 17, 2001 | Humberto<br>September 26, 2001 | Juliette<br>September 27, 2001 | Olga<br>November 28, 2001 |
| Alma<br>May 30, 2002 | Sinluka<br>September 5, 2002 | Kyle<br>September 27, 2002 | Lili<br>October 2, 2002 |

Each image has a pixel size of 400 m and has dimensions of 100 km × 100 km (© CSA).

the atmospheric boundary layer, their existence in tropical cyclones has relevance to both the heat engine that drives the vortex and the numerical models used to forecast subsequent development.

Another SAR observation of importance to our understanding of tropical cyclone dynamics is the occurrence and structure of high wind speed incursions into the cyclone's eye. These incursions are probably the result of mesovortices rotating with the cyclone's mean wind or Rossby waves propagating around the cyclone's center (Schubert *et al.* [37]). The catalogue of tropical cyclone eyes acquired to date (Fig. 10) shows a range of asymmetries and wave patterns in the surface wind in the eye region. These structures may be objectively analyzed in terms of their ellipticity and the presence of waves around the eye wall and incursions of higher winds into the eye (Du and Vachon [38]), perhaps leading to a better understanding of tropical cyclone eye dynamics and its role in vortex intensification.

"Hurricane Watch" (http://www.ccrs.nrcan.gc.ca/ccrs/rd/apps/marine/hurrican/watch_e.html), an operational initiative of the CSA, is routinely acquiring Radarsat-1 images of tropical cyclones and is providing acquisition plans to partners in order to better coordinate data collection activities such as reconnaissance flights. This new information will allow better understanding of the contributions that SAR-derived surface wind data can make towards identifying storm intensity, asymmetry, and other important characteristics.

### 2.3 Macroscale phenomena

Now we will examine SAR's ability to sense the sea surface footprints of macro (i.e., synoptic) scale fronts and extratropical cyclones. SAR provides exquisite details of the substructure of each phenomenon. We argue that those simulating, modeling, and operationally forecasting synoptic scale marine meteorology should consider SAR as a useful instrument for verification and analysis purposes.

#### 2.3.1 Fronts

Synoptic scale fronts are air mass boundaries that have collapsed down to near-zero order discontinuities in wind direction and wind speed. They are often accompanied by a surface wind maximum along and just ahead of the front (i.e., the prefrontal jet) (Carlson [39]). Thus, the SAR signature of a front most often appears as a sharp gradient in SAR intensity (e.g., Figs 5, 11, and 12). The look-angle dependence of SAR NRCS can either enhance or diminish this gradient depending on the relative orientations of the pre- and postfrontal winds to the look direction (Young *et al.* [40]).

The existence of a frontal signature implies both the existence of a front, prefrontal jet, and the lower tropospheric structures associated with them. Thus, a frontal inversion would be expected to extend from the surface front up over the prefrontal jet for a warm front and up and away from the prefrontal jet for a cold front (e.g., Young *et al.* [40]; and Fig. 13). The SAR signature of some cold fronts are marked by mesoscale vortices and/or are lobed and clefted by gravity current surges (cusps) as in Figs 5, 11 and 12 (e.g., Lee and Wilhelmson [34]), permitting them to be distinguished from the typically smoother warm fronts (e.g., Fig. 12). In addition to vortices and cusps, a wide variety of mesoscale features as well as microscale features are often observed in the vicinity SAR-detected fronts (e.g., Figs 5, 11, and 12; Young *et al.* [40]).

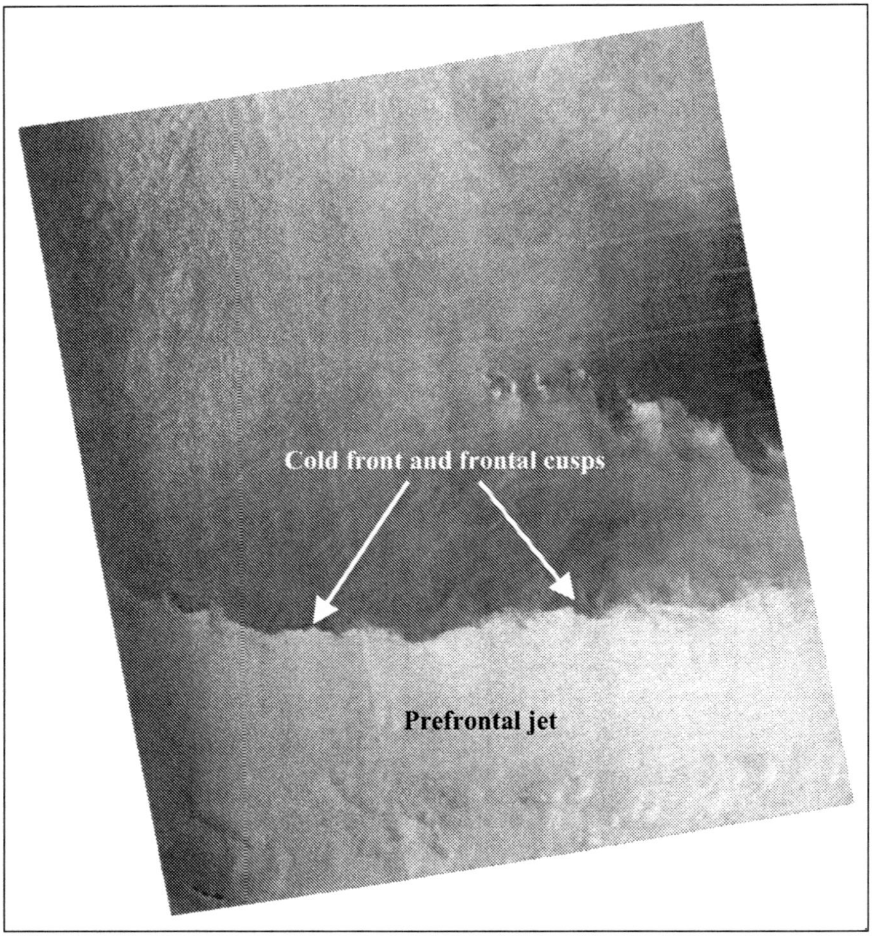


Figure 11: Radarsat-1 SAR image depicting the signature of a cold front, frontal cusps, and the prefrontal jet. The 300 m pixel image is approximately 500 km × 415 km. The image was acquired at C-band, horizontal polarization, over the Bering Sea at 0557 UTC on February 2, 2000. The top of the figure is directed towards 000°T (Provided courtesy of JHUAPL, © CSA).

### 2.3.2 Extratropical cyclones

Extratropical cyclones are typically reflected in SAR images by their impact on the associated fronts and prefrontal jets. The SAR signatures of the life cycle stages of extratropical cyclones closely resemble those structures revealed in traditional *in situ* and remote sensing analyses (e.g., Young *et al.* [40]).

An incipient cyclone appears as a kink in the front/jet signature, generally that of a cold or stationary front with a prefrontal jet on its warm side. As the cyclone

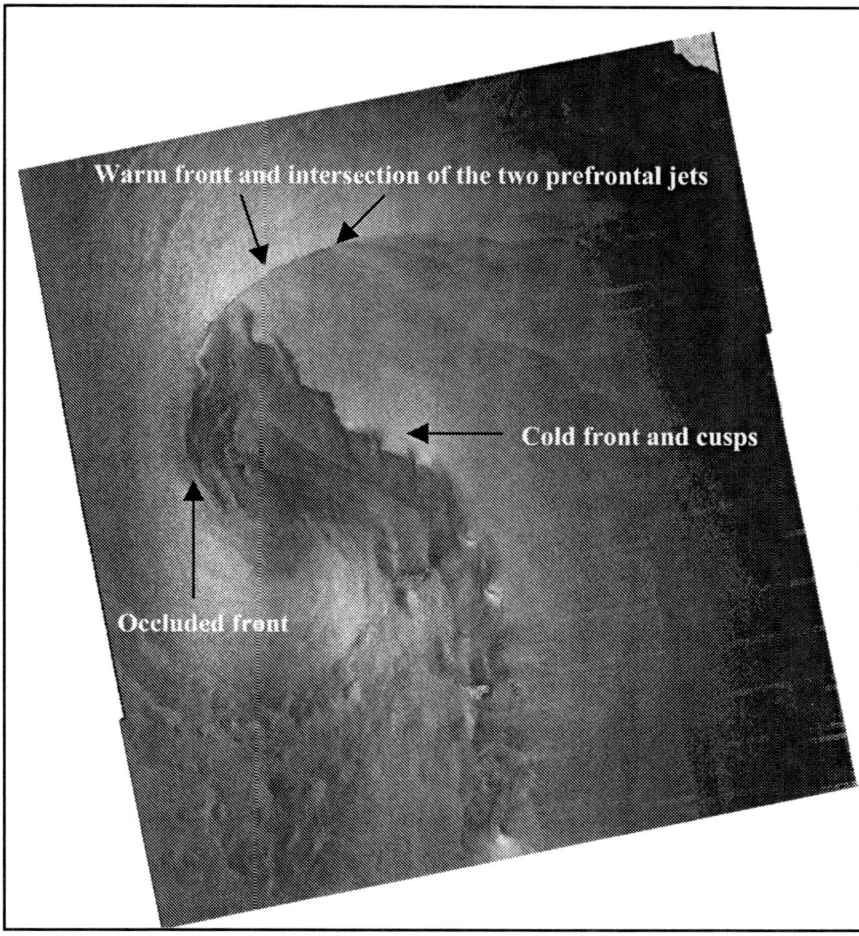


Figure 12: Radarsat-1 SAR image depicting the signature of the frontal seclusion stage of a mid-latitude cyclone. The 300 m pixel image is approximate 500 km × 415 km. The image was acquired at C-band, horizontal polarization, over the Bering Sea at 1819 UTC on December 6, 2000. The top of the figure is directed towards 000°T (Provided courtesy of JHUAPL, © CSA).

matures, the kink amplifies and the section of the front to its east develops warm frontal characteristics, including a relocation of the prefrontal jet to its cold side. Thus, a mature cyclone will exhibit two prefrontal jets with the one ahead of the cold front intersecting that ahead of the warm front near the cyclone's center. There is often a distinct minimum in the near-surface wind speed and the SAR intensity along this line of intersection (e.g., Fig. 12).

In an occluding cyclone, the prefrontal jet of the warm front extends beyond this point of intersection, sometimes for meso $\alpha$ scale distances into the cold sector.

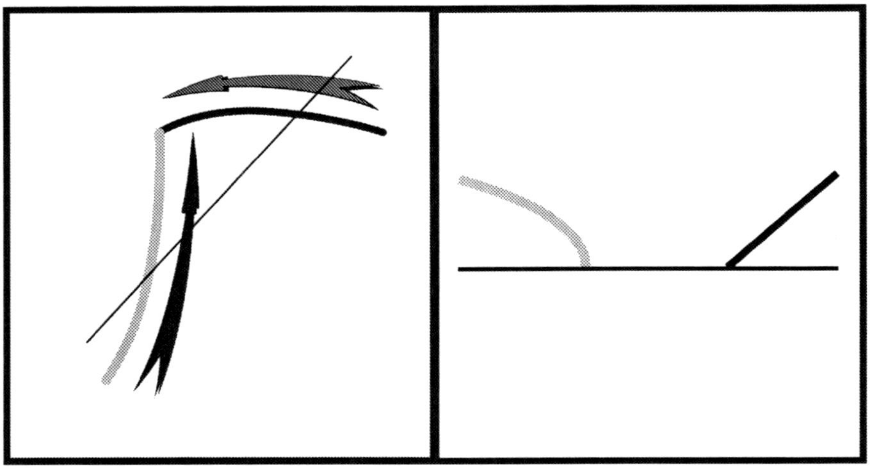

Figure 13: Schematic diagram depicting the fronts of a mature cyclone, plane view on the left and cross section on the right. The cold front is depicted as a light gray line and its prefrontal jet (the warm conveyor belt) as a dark gray arrow. The warm front is depicted as a dark gray line and its prefrontal jet (the cold conveyor belt) as a light gray arrow. The thin black line on the left indicates the orientation of the vertical cross section on the right.

The sharp gradient at the edge of this jet corresponds to the occluded front. The cyclone center lies near the end of this front. In some cyclones, the occluded front proceeds to wrap around the cyclone center forming a sharply defined circle of low winds surrounded by a frontal discontinuity and then a ring of high winds (a frontal seclusion). Despite the complicating factors of look angle and stability dependence, these signatures are often quite recognizable on SAR images (e.g., Fig. 12).

## 3 SAR-generated near-surface wind speed images

The Bragg scattering discussed in Section 1.2 is actually a first order approximation of the scattering mechanism associated with SAR images. The passage of longer ocean waves through an area illuminated by a radar, the presence of short steep waves, and foam can complicate the situation. It can be asserted that if we understood the hydrodynamic response of the ocean surface to the surface wind stress and could specify the surface structure, it would be possible to theoretically predict the expected ocean backscattering NRCS. However, in practice, the geophysical model function (GMF), the relationship between near-surface wind speed and direction to NRCS, is empirically determined.

Since many of the recently flown SARs (e.g., ERS-1 and 2, Radarsat-1, and Envisat; see Section 4 for more details) operate at C-band (approximately 5 cm in

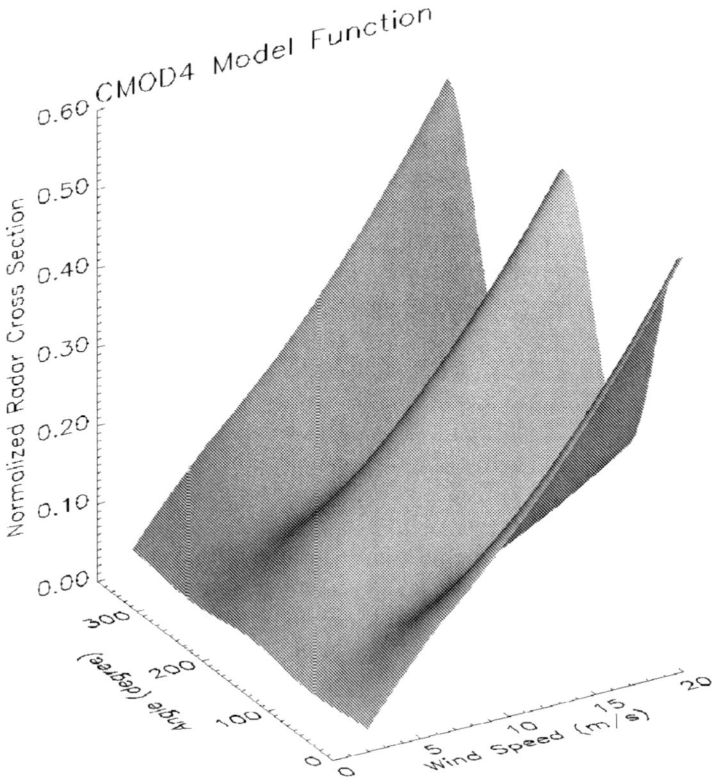


Figure 14: The CMOD4 GMF relating near-surface wind speed and direction with respect to the radar to NRCS at 25 degrees incidence.

wavelength) a considerable degree of attention has been devoted to specifying the GMF at this frequency. GMFs generally have the canonical form:

$$\sigma_0 = A(\theta)U^{\gamma(\theta)}[1 + B(\theta,\, U)\cos\varphi + C(\theta,\, U)\cos 2\varphi], \tag{1}$$

where $\sigma_0$ is the NRCS, $U$ is the near-surface wind speed, $\varphi$ is the relative angle between the wind direction and the radar look angle, $\theta$ is the local radar incident angle, and $A$, $B$, $C$, and $\gamma$ represent model parameters dependent on the incident angle and the wind speed. The dominant features of this function are that NRCS increases with near-surface wind speed, decreases with incident angle, and is a harmonic function of the angle between the wind direction and the radar look direction. Different empirical relationships for the GMF exist, but with minor variations they take on the form above.

In 1997, Stoffelen and Anderson [41] proposed the CMOD4 model function for C-band vertical polarization backscatter. Figure 14 is a representation of the CMOD4 model function for 25 degrees incidence. Recently, a revised version of this model function, CMOD5, has been proposed (Hersbach [42]).

The CMOD4 and CMOD5 GMFs yield similar wind speed retrievals for wind speeds less than about $20\,\mathrm{m\,s^{-1}}$. CMOD5 was developed to improve retrievals, particularly at higher wind speeds. The systematic evaluation and validation of SAR winds retrieved using CMOD5 is an active area of research.

Figure 14 makes clear the dilemma faced in wind retrievals using radar. The measurement of an NRCS would represent a horizontal plane slicing through the GMF. The intersection of this plane with the function would represent all the near-surface wind speed and direction pairs consistent with the measured NRCS. While a near-surface wind speed and direction produces a single NRCS, a single NRCS is associated with a large number of near-surface wind speed and direction pairs. The inversion is not unique.

Conventional radar scatterometry alleviates this problem by measuring NRCS of the ocean surface from a number of different aspect angles and/or polarizations. These additional measurements reduce the possible near-surface wind speed and direction solutions to a handful. The correct pair can usually be deduced by estimating the most likely pair from statistical considerations or from considerations of the continuity of the wind field.

A SAR typically measures the ocean surface NRCS at only a single geometry. It is, therefore, not possible to infer a near-surface wind speed and direction. If, however, we have an independent estimate of near-surface wind direction, near-surface wind speed can be inferred. This is the approach taken to convert SAR images into near-surface wind speed images.

The question of wind direction, however, is left begging. From where do the wind directions for the wind speed inversion come? There are two primary approaches. The first is to use wind direction from numerical weather models interpolated down to each SAR image pixel (e.g., Monaldo *et al.* [9]). As mentioned in Section 2.1, the second approach is to use linear features in the SAR images to estimate the wind direction (e.g., Wackerman *et al.* [43]; Fetterer *et al.* [44]; Lehner *et al.* [45]; Horstmann *et al.* [46]).

Using wind directions from numerical weather models or from linear features in the SAR image offer both important advantages and disadvantages. Model directions are always available and provide physically realistic variations in wind directions. However, models can have coarse resolution and miss or slightly displace in space and time important wind field features.

Linear features associated with the wind are not always apparent in SAR images and, as discussed in Section 2.1, are not always aligned with the near-surface wind vector and can be confused with each other. However, when they are aligned with the near-surface wind vector, linear features in the SAR images often reveal wind direction variability not resolved by numerical models. Perhaps the best approach will eventually prove to be the combination of directions from high-resolution wind models adjusted on the basis of linear features in the SAR image.

Another factor complicating current efforts to create SAR-derived near-surface wind speed images is that early GMFs were built to accommodate SARs transmitting and receiving at vertical polarization (e.g., ERS-1 and 2). Radarsat-1 transmits and receives at horizontal polarization. This means that the GMFs developed for

ERS-1 and 2 probably could not be readily applied to Radarsat-1 wind retrievals. Thompson and Beal [47], Vachon and Dobson [48], Horstmann *et al.* [49], and Wackerman *et al.* [50] have developed GMF variations to accommodate horizontal polarization.

### 3.1 Alaska SAR demonstration

Starting in 1999, the Alaska SAR Demonstration Project (Monaldo [51]) demonstrated the ability to produce near-surface wind speed images in near real time from Radarsat-1 SAR images. Radarsat-1 SAR data are downloaded to the Alaska SAR Facility (ASF) in Fairbanks, Alaska, when Radarsat-1 is in the reception region. ASF processes the images into calibrated NRCS images and transmits the data electronically to the National Environmental Satellite, Data, and Information Service in Camp Springs, MD. From there the data are converted to near-surface wind speed using two separate approaches. One is to use model directions from the Naval Operational Global Atmospheric Prediction System (NOGAPS) interpolated down to each image pixel. The result is near-surface wind speed images at subkilometer resolution. The second approach is to divide the SAR images into $25 \times 25$ km squares. From each square wind direction is inferred from linear features and a near-surface wind speed is retrieved (e.g., Wackerman *et al.* [43]). In the early months of the Alaska SAR Demonstration it took 5–6 h to go from acquisition at the satellite to the posting of near-surface wind speed images on the World Wide Web. Increases in computing power have reduced this data latency to 3 h and sometimes less.

Near-surface wind speeds retrieved intelligently using both wind direction methods typically agree with corresponding National Data Buoy Center (NDBC) buoys measurements to within better than $2\,\mathrm{m\,s^{-1}}$. Figure 15a and c shows examples of a SAR-derived near-surface wind speed image produced as part of the Alaska SAR Demonstration. The image covers a portion of the Aleutian Island Chain. The arrows in the image represent the wind vectors from the NOGAPS model. The retrieved high-resolution near-surface wind speeds are gray-scale coded. The land areas are shown as a shaded relief map. Note that the wind directions are dominantly from the northwest. As the wind passes over the topography of the Aleutian Islands, it is intensified into gap flows. Of particular interest in this case are the von Kármán vortices that are shed as the wind flow is disrupted by the Pogromni volcano.

The reader is directed to the Alaska SAR Demonstration website (http://fermi.jhuapl.edu/sar/stormwatch/index.html) to view other SAR-derived near-surface wind speed images.

## 4 SAR meteorology: a historical perspective and a look into the future

The history, current status, and future prospect of scientific SAR constitutes a tale of a continuing quest for wider swath, higher resolution, lower noise, better calibration,

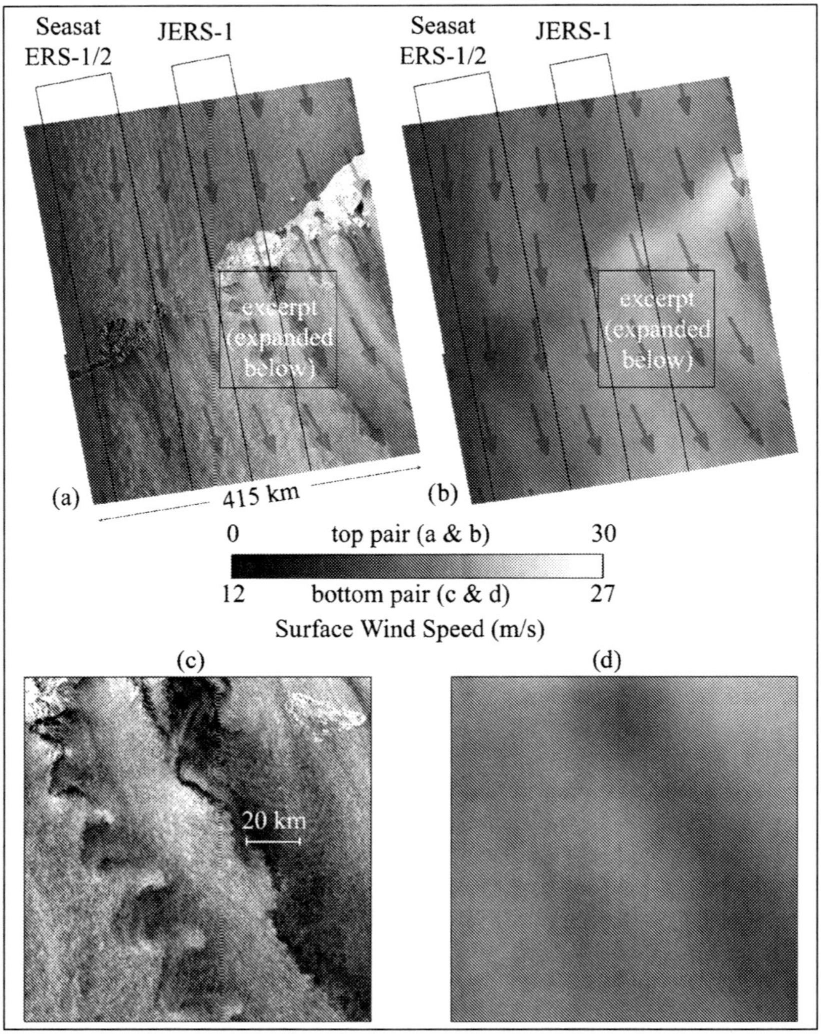


Figure 15: Radarsat-1 SAR-derived wind field depicting the signatures of von Kármán vortex streets (a and c). The pixel size is 300 m. Figure 15a has dimensions of approximately 500 km × 415 km. Also shown for comparison are the corresponding simulated 15 km pixel QuikScat wind fields (b and d). (c) and (d) are magnified ×3 and gray-scale enhanced ×2. The gray vector field is the mean ambient wind from NOGAPS. More limited swath widths of other SARs are also shown.

more accurate algorithms, and quicker delivery of targeted products to specific user communities, in particular, the operational meteorological community. In addition to the technical and scientific problems, difficult political issues must be addressed: a viable SAR global meteorological network must contain a guarantee of reliable and inexpensive access to a coordinated database from the array of internationally sponsored SARs, both currently operating and in the planning stage.

More than a quarter century has elapsed since the U.S. Seasat, containing the first civilian SAR launched solely for scientific purposes, provided the first exciting and provocative high-resolution radar images of the ocean from space (Beal *et al.* [1]). The Seasat (L-band, ~20 cm Bragg interaction wavelength, 100 km swath) SAR operated for only about 100 days from 4 July to 10 October 1978 before it failed, but the spatial patterns in its ocean surface images revealed a rich variety of physical processes that was, for the most part, quite unexpected in the scientific community. Even the first crude optically processed (with lenses instead of computers) images clearly showed ocean and atmospheric internal wave patterns, tropical storm cells, Gulf Stream signatures, spatially evolving wind-generated waves, and many other phenomena of potential interest to both oceanographers and meteorologists. The Seasat SAR, however, was for the most part uncalibrated.

In the wake of this short burst of activity in the late 1970s, no other civilian free-flying (nonshuttle) SAR was launched for more than a decade. Then, in 1991, after several years of design and preparation, ESA and the Japanese Space Agency launched ERS-1 and JERS-1, containing C-band (~5 cm interaction, 100 km swath) and L-band (~20 cm interaction, 75 km swath) SARs, respectively. ERS-1 provided the main source of high quality oceanographic SAR images during the first half of the 1990s. The JERS-1 SAR unfortunately was seriously handicapped by excessive ambiguities (ghosts) in its images originating from a faulty antenna, greatly reducing its value as a calibrated scientific instrument. The ERS-1 SAR and its identical successor, ERS-2 in 1995, provided for the first time carefully calibrated and stable instruments from which quantitative radar backscatter, in combination with an appropriate geophysical algorithm, including an estimate of the local wind direction, could yield accurate values of the surface wind magnitude at subkilometer scales, a feat not possible with any other sensor, nor indeed even easily modeled.

In 1995, another major step in the evolution of SAR was taken with the launch of the Canadian Radarsat-1. Radarsat-1 contained the first "Scan-SAR", a sophisticated improvement over all previous conventional SARs that allowed much wider swaths (>400 km) by coherently combining the returns of several antenna beams. Unfortunately the Radarsat-1 ScanSAR had an offsetting liability: an engineering oversight that allowed a nonlinear (scene-dependent) instrument transfer function, thus precluding the possibility of accurate calibration. Even so, Radarsat-1, with its 400–500 km swath width and international accessibility, has provided compelling images of high-resolution wind patterns over the ocean as demonstrated in Sections 2 and 3 above. Figure 15a and c shows just one of hundreds of such examples in which nearly periodic von Kármán vortex streets are shed from a volcanic peak in the central Aleutian Islands. Also shown for reference are the corresponding swath widths of some of the earlier SARs. The spatial detail revealed in these vortices exceed by at least an order of magnitude that which is possible with conventional scatterometers such as NSCAT (Liu *et al.* [52]) and QuikScat (Draper and Long [53]), simulated in Fig. 15b and d by scaling the SAR resolution down to 15 km.

In the latter half of 2002 and early 2003, wide swath images from the European Envisat "Advanced SAR" (ASAR) began to become available. Early indications are that most of the engineering problems associated with the Radarsat-1 ScanSAR have been largely overcome in the Envisat wide-swath ScanSAR modes. With respect to Radarsat-1, the Envisat ScanSAR antenna beam corrections are more precise, the radar system dynamic range is wider and more linear, and more attention has been given to the absolute calibration of its wide-swath modes. As a consequence, the performance of the ScanSAR itself appears finally to be only a minor source of error in the determination of near-surface wind speed. Other error sources result from uncertainties in: (i) the backscatter-to-wind-speed relationship (especially at winds higher than $\sim 15\,\mathrm{m\,s^{-1}}$) and (ii) the initial wind direction estimate (as discussed in Section 3) are now dominant. As more data from Envisat and future SARs such as Radarsat-2 and the Japanese ALOS are collected, the first error will gradually be reduced to acceptable levels. But reduction of the second, which under some circumstances, e.g., in the vicinity of fronts and within small-scale vortices such as those seen in Fig. 15c, can produce wind magnitude errors of a factor of two, will require considerable sophistication in the processing strategy. As discussed in Section 3, reduction of these kinds of errors will require blending of information both from high-resolution forecast models and from the SAR images itself.

Clearly substantial progress has been made, especially in the past decade, toward achieving well-calibrated (therefore *scientifically* viable) 400–500 km wide-swath SARs. The next step to *operational* viability will require a concerted international effort to coordinate multiple ScanSAR satellites, effectively achieving swath widths of 1200–1500 km. Such effective swath widths would for the first time allow twice daily coverage of most of the world's oceans. Figure 16 shows in graphic form one single measure of progress over the past 25 years. One can see from the graph that

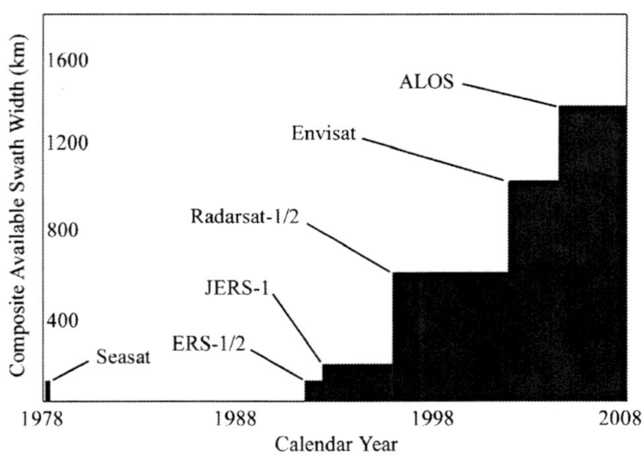


Figure 16: Composite available SAR swath widths as a function of calendar year, extrapolated past 2003.

a triad of synchronized satellites consisting of the European Envisat, the Canadian Radarsat-2, and the Japanese ALOS, each flown in identical orbits trailing the others by a third and two-thirds of an orbit (~33 and 66 min) could produce a total swath width of ~1400 km. Unfortunately, the triad will not be synchronized and, in any case, data distribution policies are not yet firmly established.

Nevertheless, it is easy to be optimistic about the future. Government reluctance to freely disseminate high-resolution SAR images is an outdated legacy arising from its value for military intelligence gathering, but high-resolution (subkilometer) wind fields are at least two orders of magnitude removed from any useful intelligence mode. So why should SAR wind fields be treated any differently from other satellite wind fields, such as NSCAT, QuikScat, or WindSAT? For example, as this section is being written, QuikScat winds are delivered to the public domain several times daily through a World Wide Web link on the home page of the NDBC (e.g., for the NDBC buoy 46035: http://www.ndbc.noaa.gov/quikscat.phtml?station=46035). It takes little imagination to see that one simple additional click on the QuikScat wind field could link to the concurrent (but much higher resolution) SAR wind field. It is the hope of the authors that public safety will eventually trump historical inertia.

## Acknowledgments

The authors are indebted to Drs. Kristina Katsaros, Susanne Lehner, and Nathaniel Winstead for the valuable input they provided during the preparation of this chapter. This work was funded by Office of Naval Research grants N00014-03-WR-20329, N00014-04-WR-20365, N00014-04-10539, and N00014-05-WR-20319; National Science Foundation grant ATM-0240869; and National Oceanic and Atmospheric Administration Office of Research and Applications contract N00024-03-D-6606.

## References

[1] Beal, R.C., DeLeonibus, P.S. & Katz, I. (eds). *Spaceborne Synthetic Aperture Radar for Oceanography*, The Johns Hopkins University Press: Baltimore and London, 1981.

[2] Orlanski, I., Rationale subdivision of scales for atmospheric processes. *Bulletin of the American Meteorological Society*, **56**, pp. 527–530, 1975.

[3] Stull, R.B., *An Introduction to Boundary Layer Meteorology*, Kluwer Academic Publishers: Dordrecht, Boston, and London, 1988.

[4] Wright, J., Backscattering from capillary waves with application to sea clutter. *IEEE Transactions on Antennas and Propagation*, **14**, pp. 749–754, 1966.

[5] Melsheimer, C., Gade, M. & Alpers, W., Multi-frequency/multi-polarization radar signatures of rain cells over the ocean measured by SIR-C/X-SAR. *Future Trends in Remote Sensing*, ed. P. Gudmandsen, Balkema: Rotterdam, pp. 85–192, 1998.

[6] Melsheimer, C., Alpers, W. & Gade, M., Simultaneous observations of rain cells over the ocean by the synthetic aperture radar aboard the ERS satellites

and by surface-based weather radars. *Journal of Geophysical Research*, **106**, pp. 4665–4677, 2001.

[7] Mourad, P.D., Footprints of atmospheric phenomena in synthetic aperture radar images of the ocean surface—a review. *Air-Sea Exchange: Physics, Chemistry, and Dynamics*, ed. G.L. Geernaert, Kluwer Academic Publishers: Dordrecht, Boston, and London, pp. 269–290, 1999.

[8] Sikora, T.D. & Young, G.S., Wind-direction dependence of quasi-2D SAR signatures. *Proc. Int. Geoscience and Remote Sensing Symposium 2002*, IEEE, pp. 1887–1889, 2002.

[9] Monaldo, F.M., *et al.*, The SAR measurement of ocean surface winds: an overview. *Proc. Second Workshop on Coastal and Marine Applications of SAR*, European Space Agency SP-565, pp. 15–32, 2004.

[10] Sikora, T.D., Young, G.S., Beal, R.C. & Edson, J.B., Use of spaceborne synthetic aperture radar imagery of the sea surface in detecting the presence and structure of the convective marine atmospheric boundary layer. *Monthly Weather Review*, **123**, pp. 3623–3632, 1995.

[11] Zecchetto, S., Trivero, P., Fiscella, B. & Pavese, P., Wind stress structure in the unstable marine surface layer detected by SAR. *Boundary-Layer Meteorology*, **86**, pp. 1–28, 1998.

[12] Babin, S.M., Sikora, T.D. & Winstead, N.S., A case study of satellite synthetic aperture radar signatures of spatially evolving atmospheric convection over the western Atlantic Ocean. *Boundary-Layer Meteorology*, **106**, pp. 527–546, 2003.

[13] Müller, G., Brümmer, B. & Alpers, W.R., Roll convection with an Arctic cold-air outbreak: interpretation of in situ aircraft measurements and spaceborne SAR imagery by a three-dimensional atmospheric model. *Monthly Weather Review*, **127**, pp. 363–380, 1999.

[14] Alpers, W. & Brümmer, B., Atmospheric boundary layer rolls observed by the synthetic aperture radar aboard the ERS-1 satellite. *Journal of Geophysical Research*, **99**, pp. 12613–12621, 1994.

[15] Weckwerth, T.M., Wilson, J.W., Wakimoto, R.M. & Crook, N.A., Horizontal convective rolls: determining the environmental conditions supporting their existence and characteristics. *Monthly Weather Review*, **125**, pp. 505–526, 1997.

[16] Weckwerth, T.M., Horst, T.W. & Wilson, J.W., An observational study of the evolution of horizontal convective rolls. *Monthly Weather Review*, **127**, pp. 2160–2179, 1999.

[17] Brown, R.A., On the inflection point instability of a stratified Ekman boundary layer. *Journal of the Atmospheric Sciences*, **29**, pp. 850–859, 1972.

[18] Stensrud, D.J. & Shirer, H.N., Development of boundary layer rolls from dynamic instabilities. *Journal of the Atmospheric Sciences*, **45**, pp. 1007–1019, 1988.

[19] Vachon, P.W., Johannessen, J.A. & Browne, D.P., ERS-1 SAR signatures of atmospheric gravity waves. *IEEE Transactions on Geoscience and Remote Sensing*, **33**, pp. 1014–1025, 1995.

[20] Winstead, N.S., Sikora, T.D., Thompson, D.R. & Mourad, P.D., Direct influence of gravity waves on surface-layer stress during a cold air outbreak, as shown by synthetic aperture radar. *Monthly Weather Review*, **130**, pp. 2764–2776, 2002.

[21] Young, G.S., Perugini, S.M. & Fairall, C.W., Convective wakes in the equatorial western Pacific during TOGA. *Monthly Weather Review*, **123**, pp. 110–123, 1995.

[22] Atlas, D., Origin of storm footprints on the sea seen by synthetic aperture radar. *Science*, **266**, pp. 1364–1366, 1994.

[23] Gaynor, J.E. & Mandics, P.A., Analysis of the tropical marine boundary layer during GATE using acoustic sounder data. *Monthly Weather Review*, **106**, pp. 223–232, 1978.

[24] Heinemann, G. & Claud, C., Meeting summary: report of a workshop on "theoretical and observational studies of polar lows" of the European Geophysical Society Polar Lows Working Group. *Bulletin of the American Meteorological Society*, **78**, pp. 2643–2658, 1997.

[25] Bresch, J.F., Reed, R.J. & Albright, M.D., A polar-low development over the Bering Sea: analysis, numerical simulation, and sensitivity experiments. *Monthly Weather Review*, **125**, pp. 3109–3130, 1997.

[26] Nielsen, N.W., An early-autumn polar low formation over the Norwegian Sea. *Journal of Geophysical Research*, **102**, pp. 13955–13973, 1997.

[27] Miner, T., Sousounis, P.J., Wallman, J. & Mann, G., Hurricane Huron. *Bulletin of the American Meteorological Society*, **81**, pp. 223–236, 2000.

[28] Chunchuzov, I., Vachon, P.W. & Ramsay, B., Detection and characterization of mesoscale cyclones in RADARSAT synthetic aperture radar images of the Labrador Sea. *Canadian Journal of Remote Sensing*, **26**, pp. 213–230, 2000.

[29] Sikora, T.D., Friedman, K.S., Pichel, W.G. & Clemente-Colon, P., Synthetic aperture radar as a tool for investigating polar mesoscale cyclones. *Weather and Forecasting*, **15**, pp. 745–758, 2000.

[30] Friedman, K.S., Sikora, T.D., Pichel, W.G., Clemente-Colon, P. & Hufford, G., Using spaceborne synthetic aperture radar to improve marine surface analyses. *Weather and Forecasting*, **16**, pp. 270–276, 2001.

[31] Bond, N.A. & Shapiro, M.A., Polar lows over the Gulf of Alaska in conditions of reverse shear. *Monthly Weather Review*, **119**, pp. 551–572, 1991.

[32] Douglas, M.W., Shapiro, M.A., Fedor, L.S. & Saukkonen, L., Research aircraft observations of a polar low at the east Greenland ice edge. *Monthly Weather Review*, **123**, pp. 5–15, 1995.

[33] Montgomery, M.T. & Farrell, B.F., Polar low dynamics. *Journal of the Atmospheric Sciences*, **49**, pp. 2484–2505, 1992.

[34] Lee, B.D. & Wilhelmson, R.B., The numerical simulation of non-supercell tornadogenesis. Part I: Initiation and evolution of pretornadic misocylone circulations along a dry outflow boundary. *Journal of the Atmospheric Sciences*, **54**, pp. 32–60, 1997.

[35] Katsaros, K.B., Vachon, P.W., Black, P.G., Dodge P.P. & Uhlhorn, E.W., Wind fields from SAR: could they improve our understanding of storm dynamics? *Johns Hopkins APL Technical Digest*, **21**, pp. 86–93, 2000.

[36] Katsaros, K.B., Vachon, P.W., Liu, W.T. & Black, P.G., Microwave remote sensing of tropical cyclones from space. *Journal of Oceanography*, **58**, pp. 137–151, 2002.

[37] Schubert, W.H., Montgomery, M.T., Taft, R.K., Guinn, T.A., Fulton, S.R., Kossin, J.P. & Edwards, J.P., Polygonal eyewalls, asymmetric eye contraction, and potential vorticity mixing in hurricanes. *Journal of the Atmospheric Sciences*, **56**, pp. 1197–1223, 1999.

[38] Du, Y. & Vachon, P.W., Characterization of hurricane eyes in RADARSAT-1 images with wavelet analysis. *Canadian Journal of Remote Sensing*, **29**, pp. 491–498, 2003.

[39] Carlson, T.N., Airflow through mid-latitude cyclones and the comma cloud pattern. *Monthly Weather Review*, **108**, pp. 1498–1509, 1980.

[40] Young, G.S., Sikora, T.D. & Winstead, N.S., Use of synthetic aperture radar in fine-scale surface analysis of synoptic-scale fronts at sea. *Weather and Forecasting*, **20**, pp. 311–327, 2005.

[41] Stoffelen, A. & Anderson, D., Scatterometer data interpretation: measurement and inversion. *Journal of Atmospheric and Oceanic Technology*, **14**, pp. 1298–1313, 1997.

[42] Hersbach, H., CMOD5: an improved geophysical model function for ERS C-band scatterometry. *Technical Memorandum 395*, European Centre for Medium-Range Weather Forecasts: Shinfield Park, Reading, 2003.

[43] Wackerman, C.C., Rufenach, C.L., Shuchman, R.A., Johannessen, J.A. & Davidson, K.L., Wind vector retrieval using ERS-1 synthetic aperture radar imagery. *IEEE Transactions on Geoscience and Remote Sensing*, **34**, pp. 1343–1352, 1996.

[44] Fetterer, F., Gineris, D. & Wackerman, C.C., Validating a scatterometer wind algorithm for ERS-1 SAR. *IEEE Transactions on Geoscience and Remote Sensing*, **36**, pp. 479–492, 1998.

[45] Lehner, S., Horstmann, J., Koch, W. & Rosenthal, W., Mesoscale wind measurements using recalibrated ERS SAR images. *Journal of Geophysical Research*, **103**, pp. 7847–7856, 1998.

[46] Horstmann, J., Koch, W., Lehner, S. & Tonboe, R., Ocean winds from RADARSAT-1 ScanSAR. *Canadian Journal of Remote Sensing*, **28**, pp. 524–533, 2002.

[47] Thompson, D.R. & Beal, R.C., Mapping high-resolution wind fields using synthetic aperture radar. *Johns Hopkins APL Technical Digest*, **21**, pp. 58–67, 2000.

[48] Vachon, P.W. & Dobson, F.W., Wind retrieval from RADARSAT SAR images: selection of a suitable C-band HH polarization wind retrieval model. *Canadian Journal of Remote Sensing*, **26**, pp. 306–313, 2000.

[49] Horstmann, J., Koch, W., Lehner, S. & Tonboe, R., Wind retrieval over the ocean using synthetic aperture radar with C-band HH polarization, *IEEE Transactions on Geoscience and Remote Sensing*, **38**, pp. 2122–2131, 2000.

[50] Wackerman, C.C., Clemente-Colón, P., Pichel, W.G. & Li, X., A two-scale model to predict C-band VV and HH normalized radar cross section values over the ocean. *Canadian Journal of Remote Sensing*, **28**, pp. 367–384, 2002.

[51] Monaldo, F., The Alaska SAR demonstration and near-real-time synthetic aperture radar winds. *Johns Hopkins APL Technical Digest*, **21**, pp. 75–79, 2000.

[52] Liu, W.T., Tang, W. & Polito, P.S., NASA scatterometer provides global ocean-surface wind fields with more structures than numerical weather prediction. *Geophysical Research Letters*, **25**, pp. 761–764. 1998.

[53] Draper, D.W. & Long, D.G., An assessment of SeaWinds on QuikSCAT wind retrieval. *Journal of Geophysical Research*, **107**, pp. 3212–3226, 2002.

# CHAPTER 5

# Relationship between oceanic energy fluxes and surface winds during tropical cyclone passage

L.K. Shay[1] & S.D. Jacob[2]

[1]*Rosenstiel School of Marine and Atmospheric Science, Division of Meteorology and Physical Oceanography, University of Miami, Miami, FL 33149, USA.*

[2]*Goddard Earth Science and Technology Center, University of Maryland Baltimore County, Baltimore, MD 21250, USA.*

## Abstract

Based on velocity profiles acquired from the deployment of airborne expendable current profilers (AXCPs) during the passage of Hurricane Gilbert, WKBJ-scaled vertical wavenumber spectra were estimated and compared to internal wave spectra. For the stratification, observed kinetic energies were typically up to 10 times larger than the background spectral values. Clockwise (CW) rotating spectral energy densities exceeded the counterclockwise (CCW) rotating densities in 83% of the profiles by a factor of four. The downward vertical energy fluxes ranged from 0.5 to 10 ergs cm$^{-2}$ s$^{-1}$, and the average value was ~2 ergs cm$^{-2}$ s$^{-1}$, equaling those found in oceanic fronts. These near-inertial energy fluxes followed a tanh dependence where the argument involves the ratio of the CW and CCW energies.

Amplitudes of the wave groups determined from scaled-velocity profiles across the top of the thermocline ranged between 15 and 50 cm s$^{-1}$. Equating 10-m winds to near-inertial energy fluxes suggested a nonlinear relationship for the drag coefficient, valid for winds up to ~38 m s$^{-1}$. Comparisons of this function to various accepted drag coefficient formulations were favorable between wind speeds of 12–20 m s$^{-1}$. For wind speeds beyond 28 m s$^{-1}$, this function leveled off, compared to monotonically increasing wind-speed dependent drag coefficient formulations. Mixed layer current simulations from a high-resolution ocean model, forced with observed winds with varying surface drag coefficients, suggest reasonable agreement to those simulated from this analytical function. Simulated currents were within ~20% to those observed from AXCPs. Understanding the relationships associated with the surface momentum flux is important for air–sea coupling physics during strong wind conditions.

## 1 Introduction

Tropical cyclone (TC) passage over an ocean represents one of the more extreme upper ocean response cases, as the ocean loses heat to the atmosphere. The most apparent effect of TC passage is the marked cooling of the sea surface temperatures (SSTs) of 1–5°C, which occurs to the right of the storm track, as shown for Hurricane Juliette in the eastern Pacific Ocean (Fig. 1). Although warm SSTs (>26°C) are required to maintain the TC, the *rightward bias* of the maximum SST cooling and mixed layer depth deepening by 20–40 m are primarily due to entrainment mixing of the cooler thermocline water with the warmer ocean mixed layer (OML) water, which accounts for about 65–80% of the observed SST decreases. These SST decreases are due to either vertical shear of the horizontal currents or surface-generated turbulence within the OML. Both mechanisms play important roles in the upper ocean thermodynamics and air–sea fluxes.

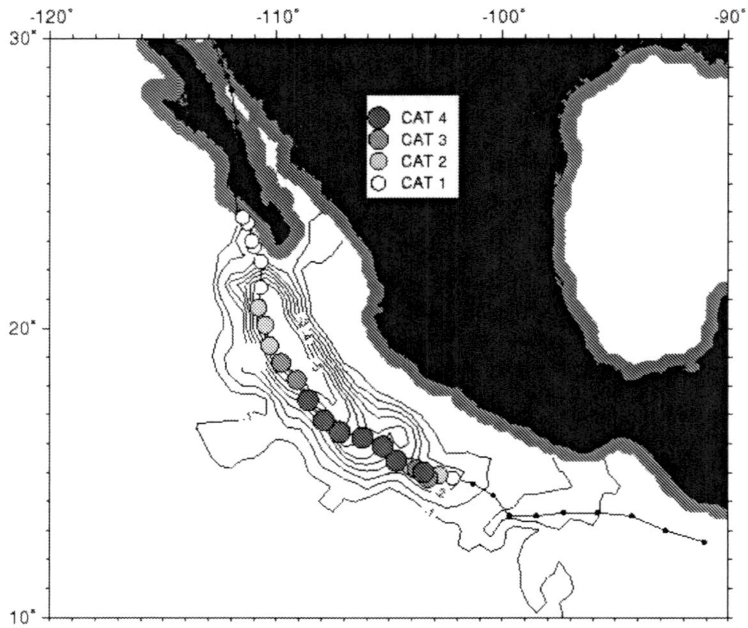


Figure 1: SST response to the passage of Hurricane Juliette in September 2001 relative to the track (black curve) and storm intensity (see legend). SST is derived from TRMM Microwave Imager (TMI), using a 75 km coastal ocean blanking.

The ocean's momentum response is classified into two regimes: the directly forced or near-field; and the evolving three-dimensional wake or far-field (Shay [1]). In the near-field, the cyclonically rotating wind field of a TC causes OML currents of about 1–1.5 m $s^{-1}$ to diverge from the storm track, starting within one-quarter of an inertial wavelength behind the eye, where the inertial wavelength ($\Lambda$) is defined as the product of the storm translation speed ($U_h$) and the local inertial period (IP). This divergence causes the upwelling of cooler water underneath the storm track, thereby decreasing the OML depth. Over the next one-half inertial cycle, OML currents converge towards the storm track, causing an increase in the OML depth as warmer water is downwelled into the thermocline. This alternating cycle of upwelling and downwelling of the isotherms (and isopycnals) occurs over distances of $\Lambda$ and establishes horizontal pressure gradients that couple the OML to the thermocline as part of a spreading three-dimensional wake.

The near-inertial response represents one of the more dominant components that have been observed and modeled during TC passage. Within this framework, near-inertial motions are characterized by large horizontal and vertical wavelengths (Geisler [2]) due to oscillations in the vertical current structure by low-order baroclinic modes (Gill [3]; Kundu and Thomson [4]; Shay *et al.* [5]). An important aspect of these forced motions has been the frequency shift above the local Coriolis frequency by about 5–10%. This shift is necessary for the near-inertial energy to vertically propagate from the surface-intensified OML into the thermocline. However, during strong forcing events, both the rate of vertical energy flux and the level where energy accumulates have remained unresolved due to the lack of observational evidence.

From a series of airborne expendable current profilers (AXCPs) successfully deployed in TCs Josephine (1984), Norbert (1984), and Gloria (1985), as part of the Ocean Response to A Hurricane Project (ORTAH) sponsored by a consortium of oil companies, high quality current profiles were obtained in a hurricane environment for the first time (Sanford *et al.* [6]). While these baroclinic current and temperature profiles were studied observationally (Sanford *et al.* [6]), analytically (Shay *et al.* [5]), and numerically (Price *et al.* [7]), the time-evolution of the spreading three-dimensional near-inertial wave wake was not resolved with only one snapshot of measurements. Subsequently, Shay *et al.* [8, 9] isolated the near-inertial response induced by Hurricane Gilbert from AXCPs deployed before, during, and after hurricane passage. Upon removal of the geostrophically balanced currents associated with a loop current warm-core ring (LCWCR), velocity profiles were placed in a storm-based coordinate system where time represented along-track distance for a steadily moving storm (Geisler [2]). Using least squares, these profiles were modeled to isolate the near-inertial content and compared to forced, analytical simulations. A summation of the first four baroclinic modes described up to 77% of this near-inertial current variability over the first 1.7 IPs($\sim$30 h) following Gilbert.

There is a characteristic time scale for the phase of each baroclinic mode to separate from the wind-forced OML currents when the wind stress scale ($\sim$2)$R_{max}$ defined as the radius of maximum winds) is greater than the deformation radius associated with the first baroclinic mode of $O$(40 km) (Gill [3]). This time scale, which is required for a phase difference of $\pi/2$ to develop between the upper thermocline and the OML currents, increases with mode number due to decreasing phase speeds. These baroclinic time scales represent the onset of rapid energy propagation from the OML into the thermocline (Zervakis and Levine [10]). In a theoretical framework, an average energy flux from moving TCs to the internal wave field is found to have a global value of $10^{10}$ W by summing forced normal modes (Nilsson [11]). However, estimating the energy fluxes has been more problematic in strong TCs due to the lack of three-dimensional current profiles in the near-inertial wave wake. Moreover, the relationship of these fluxes to the near-surface winds has been even more difficult due to the lack of synopticity in concurrent oceanic current and atmospheric wind measurements over storm scales.

An important question emerging from studies of a strongly forced OML is the relationship between the internal wave flux and surface wind stress. Under storm conditions, considerable uncertainty remains with respect to the surface drag coefficient used to estimate the wind stress. For example, Large and Pond [12] found a monotonically increasing drag coefficient, valid in winds to 25 m s$^{-1}$. For the response to Hurricanes Norbert and Josephine, simulated currents agreed well with OML currents from Price *et al.* [7] based on Large and Pond's [12] drag coefficients extrapolated to higher winds. Buoy measurements in Hurricanes Anita, Camille, and Frederic suggest a slightly higher bias in the drag coefficient, but with the same slope up to 45 m s$^{-1}$ (WAMDI Group [13]). However, the wave-age dependent drag coefficient formulation of Donelan [14] indicates considerable variability in the surface drag between 10 and 20 m s$^{-1}$ compared to these wind-speed dependent drag coefficients. While direct methods have been used to estimate the surface drag coefficient, Smith *et al.* [15] suggested using inverse methods and constraints to determine the exchange coefficients associated with heat and momentum fluxes. This approach may be particularly important at very high winds and sea states where direct measurements at the air–sea interface are difficult to acquire in TCs. Thus, one possibility is to use the ocean's momentum response as a tracer of surface wind stress, and hence determine the drag coefficient from observations during TCs.

Accordingly, a functional dependence between the observed surface winds and wind-forced internal wave fluxes is examined here using the current profiles for Hurricane Gilbert, estimated by Shay *et al.* [8] in a first attempt to use this approach. The resulting functional relationship as well as several accepted drag coefficient methods (Large and Pond [12]; WAMDI Group [13]; Smith [16]) are then used to simulate OML currents from the Hybrid Coordinate Ocean Model (HYCOM) to assess differences in the observed and simulated OML profiles (Bleck [17]).

As shown in Fig. 2, these measurements included 76 AXCPs deployed prior, during, and subsequent to the passage of Hurricane Gilbert from September 14–19, 1988 (Shay *et al.* [8]). Given the predominance of the near-inertial wave excitation by the wind field, vertical wavenumber spectra are estimated for WKBJ-scaled profiles and compared to estimates from the internal wave spectra (Garrett and Munk [18]). Kinetic energy spectral densities are decomposed into clockwise

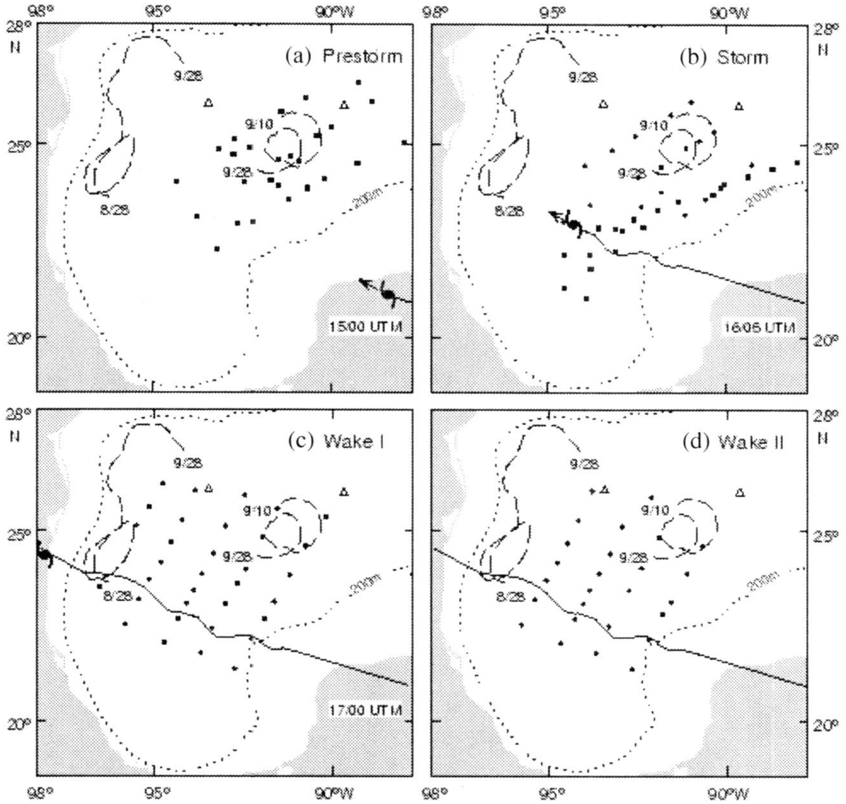


Figure 2: Locations of AXCPs (circles) and AXBTs (boxes), tracks of drifting buoys (dashed), National Data Buoy Office buoys (triangles), and Minerals Management Service moorings EE, FF, and GG (open circles in panel a) relative to the track of Hurricane Gilbert (solid) in the Gulf of Mexico from (a) September 14, (b) September 16, (c) September 17, and (d) September 19, 1988. The 200 m bottom contour is dotted (from Shay *et al.* [8]).

(CW)- and counterclockwise (CCW)-rotating components to estimate the rates of downward energy flux. That is, excess spectral energy densities (i.e. CW–CCW-rotating components) are multiplied by the group velocity and integrated over the vertical wavenumber space following Leaman [19] to determine the internal wave flux in the three-dimensional wake structure. These internal wave fluxes are equated to the observed surface (10 m) winds to derive a functional relationship. These results are placed into a framework of the surface drag coefficients found from previous studies (Large and Pond [12]; WAMDI Group [13]; Donelan [14]; Smith [16]; Alamaro [20]; Powell *et al.* [21]). The implications of this approach are summarized and similarities to characteristics of well-established drag coefficients are noted.

## 2 Observations

This section focuses on analysis of measurements made during Hurricane Gilbert (1988) as it propagated across the Gulf of Mexico. Thus, we provide surface winds, air–sea variables and velocity profiler data in estimating the kinetic energy fluxes into the thermocline from the wind-forced OML. Measurements and profiler analyses have been described elsewhere (Shay *et al.* [8, 9]).

### 2.1 Wind structure

Flight-level winds were reduced to 10 m using a boundary layer model of Powell [22] assuming a translating wind field and objectively analyzed at 8 km intervals (Fig. 3). Maximum winds were ~38 and 34 m s$^{-1}$ at radii of 60 km ($R_{\max}$) and 90 km from the storm track, respectively. The wind field was broad and complex

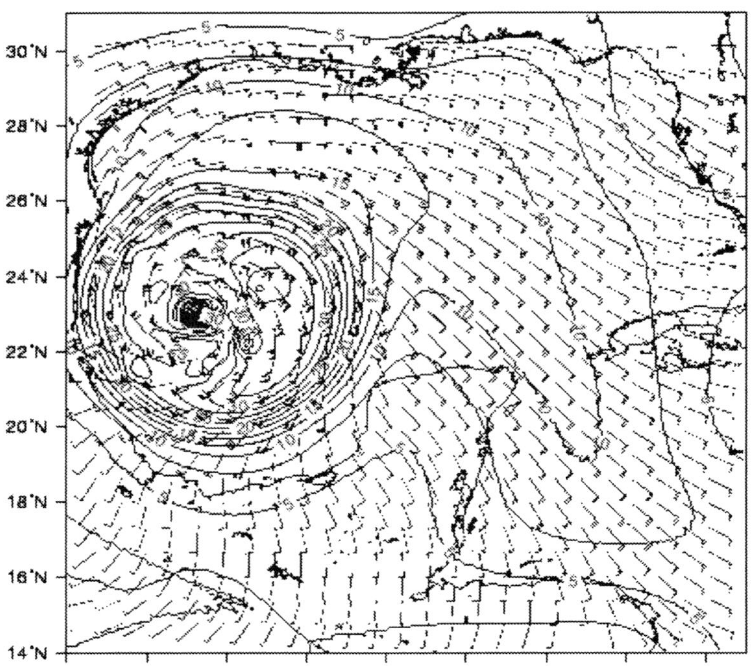


Figure 3: Surface winds (10 m) derived by combining winds reduced from flight-level, ECMWF surface analyses, and buoy observations for 06UTC September 16, 1988, observed during Hurricane Gilbert. Analyzed wind data are subsampled for every eighth data point as wind barbs (in knots) with isotachs representing wind magnitudes (m s$^{-1}$) (from Jacob *et al.* [24]).

where winds exceeded $15\,\mathrm{m\,s^{-1}}$ as far as $5R_{\mathrm{max}}$ to the right of the track over a LCWCR in the Gulf of Mexico.

This strong wind forcing field shifts the frequency of forced near-inertial motions by a few percentage above the local Coriolis frequency (Price [23]). This frequency shift is proportional to the OML Burger number, which is a function of the storm speed ($U_{\mathrm{h}}$), $R_{\mathrm{max}}$, the local Coriolis parameter ($f$) and stratification beneath the OML. In the present case, the Burger number is 0.08. Based on scaling arguments, the predicted blue shift in the near-inertial frequency is $1.04f$, which allows energy to propagate vertically from the OML into the thermocline (Shay *et al.* [9]). The predicted wind-driven ocean velocity ($V_{\mathrm{is}}$) scale, defined as

$$\tau R_{\mathrm{max}}\,(\rho_{\mathrm{o}}\,h\,U_{\mathrm{h}})^{-1},$$

is $\sim 1.1\,\mathrm{m\,s^{-1}}$. This is based upon a wind stress ($\tau$) using an extrapolated drag coefficient from Large and Pond [12], where $\rho_{\mathrm{o}}$ is the reference ocean density, and $h$ is the initial OML depth (40 m). The corresponding thermocline velocity ($V_{\mathrm{th}} = 44\,\mathrm{cm\,s^{-1}}$) is approximately twice the scaled velocity of $22\,\mathrm{cm\,s^{-1}}$, as estimated from $V_{\mathrm{is}}\,h\,b^{-1}$ where $b$ is a seasonal thermocline thickness of about 200 m suggested by observed velocity and temperature profiles.

### 2.2 Velocity profiles

Since TC Gilbert moved steadily at 290°T and at a nearly constant speed, along-track distance can be converted into time (Geisler [2]), and velocity profiles deployed at different times have an along-track position in storm coordinates relative to Gilbert's eye (22°57′N, 94°44′W). In this storm-based coordinate system, drop sites are a function of the along-track coordinate ($\Lambda$), or IP, and the cross-track coordinate $R_{\mathrm{max}}$.

Along-track velocity sections indicate a CW-rotation of the current vector relative to the storm profile, particularly in the OML (Shay *et al.* [9]). The maximum current response ($1.3\,\mathrm{m\,s^{-1}}$) normalized to the predicted wind-driven current was about 1.1 located at about $2R_{\mathrm{max}}$ which agrees with the rightward bias in the maximum response. This behavior is consistent with strong near-inertial currents that develop within the thermocline following Gilbert's passage. That is, a large fraction of the variability is associated with forced near-inertial motions and the first few baroclinic modes (i.e. low-mode response). After removal of the geostrophic current and surface wave-induced orbital velocities (Sanford *et al.* [6]), a large fraction of the current variability was associated with the near-inertial current response, and the CW-rotation of the velocity vector with depth.

### 2.3 Maximum energy levels

Maximum energy levels for the WKBJ-scaled current profiles, for Gilbert, normalized by the mixed layer depths ($h$) at each AXCP drop site, are shown in Fig. 4. Levels of energy maxima were located from the OML base (35–55 m) to about 150 m. Approximately 95% of the velocity profiles indicated maximum energy

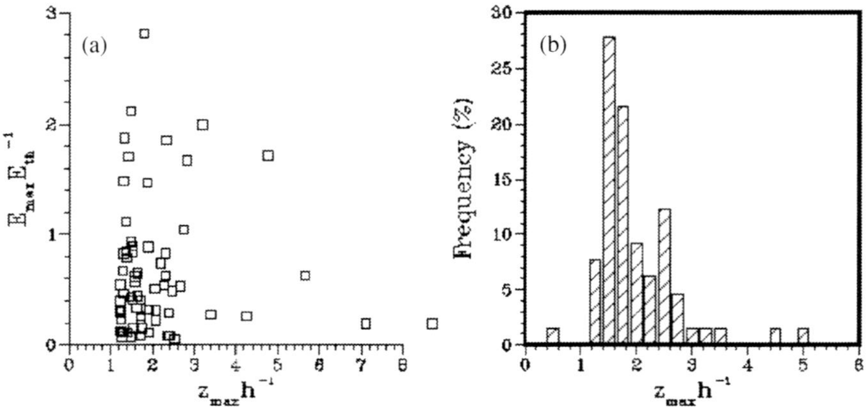


Figure 4: (a) Scatterplot and (b) histogram of levels (m) of maximum observed WKB-scaled kinetic energy scaled by the predicted thermocline kinetic energy ($E_t$) relative to the respective depth, normalized by the OML depths ($h$) at each of the current profiles for Gilbert. Note that $V_{th}$ is twice the scaled thermocline velocity, using scaling arguments from Price [23].

levels ($E_{max} \times E_{th}^{-1}$, where $E_{th}$ is estimated from the scaled thermocline velocity of $44\,cm\,s^{-1}$) from 1 to $4h$, where the average depth was $2h$ and the standard deviation was $0.8h$. The distribution tends to follow a $\chi^2$ distribution starting at about one layer depth ($h$). Since OML temperatures showed regions of upwelling (Storm and Wake 1—see Fig. 2) and downwelling (Wake 2), these depths of maximum WKB-scaled energy were representative of an energetic thermocline response to the applied atmospheric forcing. Maximum current shears coincided with the region of the largest buoyancy frequencies to 12 cph (cycles per hour), suggestive of strongly forced, near-inertial motions. These near-inertial motions reduced the bulk Richardson number to below criticality, which induced vertical mixing events across the entrainment zone (Shay *et al.* [8]). Using a diagnostic approach, Jacob *et al.* [24] found that shear-induced mixing events dominated the OML heat balance in the wake of Gilbert between 1 and $3R_{max}$ where the maximum observed OML temperature response was ~4°C.

## 3 Vertical wavenumber spectra

Velocity profiles were stretched, normalized by the buoyancy frequency, and interpolated to 3-m vertical resolution following Leaman and Sanford [25]. While the vertical wavelengths of the forced motions were large, the WKBJ-scaling allows a direct comparison to the internal wave spectra (Garrett and Munk [18]) as well as estimation of the vertical energy fluxes from the profiler data. That is, OML velocity components and energies are precluded from this calculation since the buoyancy frequency ($N$) is 0 in a well-mixed surface layer. This approach to scaling the velocity profiles provides a more realistic estimate of the energies and fluxes penetrating into the thermocline, than would be obtained by retaining OML currents.

### 3.1 Garrett–Munk comparisons

Following Leaman [19], the form of spectra is

$$\text{KE (m)} = N_o^2 M_o^{-3} 3E_o (2j_* \pi)^{-1} A(\lambda_v), \tag{1}$$

where $A(\lambda_v) = (t-1)(1+\lambda_v)^{-t}$, $\lambda_v = m(j_* \pi M_o)^{-1}$, $M_o = 0.122$ cycles km$^{-1}$, $j_* = 6$, 15, and 20, $m$ is the vertical wavenumber, $E_o = 2\pi \times 10^{-5}$ and $t = 2.5$. Within a factor of 2–3, this nondimensional energy density is a universal factor for a background internal wave continuum in the ocean, based on the thermal structure (Munk [26]). However, average values of the normalized spectral energies observed during Gilbert were 23, 11, and 9% for $j_* = 6$, 15, and 20, respectively. Thus, the energies excited by Gilbert were significantly above the spectral levels of eqn (1), for the common values of $j_*$ which represent larger wavelengths supported by the stratification and the low-baroclinic mode response.

### 3.2 Rotary spectra

Rotary vertical wavenumber spectra (Gonella [27]) of the WKBJ-scaled kinetic energy spectral densities were estimated by smoothing the 3-m velocity profiles using a Tukey window, transforming the data, and spectrally averaging to decrease leakage to adjacent wavenumber bands (Otnes and Enochson [28]). After 1.5IP following passage of Gilbert, the cross-track spectral structure encompassed the Gulf common water and the LCWCR (Fig. 5). Between $-R_{max}$ and $2R_{max}$, there was a predominance of CW-rotating motions at the low wavenumbers including the transition zone in the LCWCR regime. The transition between vertical wavenumbers became more apparent at vertical wavelengths of about 100 m. In the cross-track direction, the CW-rotating spectral peak at 500 m wavelength ($7 \times 10^{-3}$ cycles m$^{-1}$) was shifted to about 300 m wavelength ($10^{-2}$ cycles m$^{-1}$) at $1R_{max}$, and to 100–200 m wavelength ($1.5$–$3 \times 10^{-2}$ cycles m$^{-1}$) at $2R_{max}$, but was not as clearly evident at $3R_{max}$. Presumably, the distinct change, even in the higher vertical wavenumbers, may have been due to this transition zone between the Gulf common water and the LCWCR. Between 4 and $5R_{max}$, a distinguishable peak in the spectral energy densities was shifted to a vertical wavelength of 100 m. At low vertical wavenumbers (wavelengths of 500–2000 m), the WKB-scaled, CW-rotating energy dominated the CCW-rotating components. The presence of shorter wavelengths from 50 to 500 m indicated that energy input from the wind field excited a spectrum of wave modes. This is not a surprising result given the predominance of the energies in the larger vertical wavelength oscillations (low wavenumbers). This is known as the low-mode dominance of the oceanic current response excited by TCs (Price *et al.* [7]; Shay *et al.* [9]). Similarly, spectral energy densities at $2R_{max}$ in the along-track direction indicate a cascade of energy from low to high vertical wavenumbers (not shown).

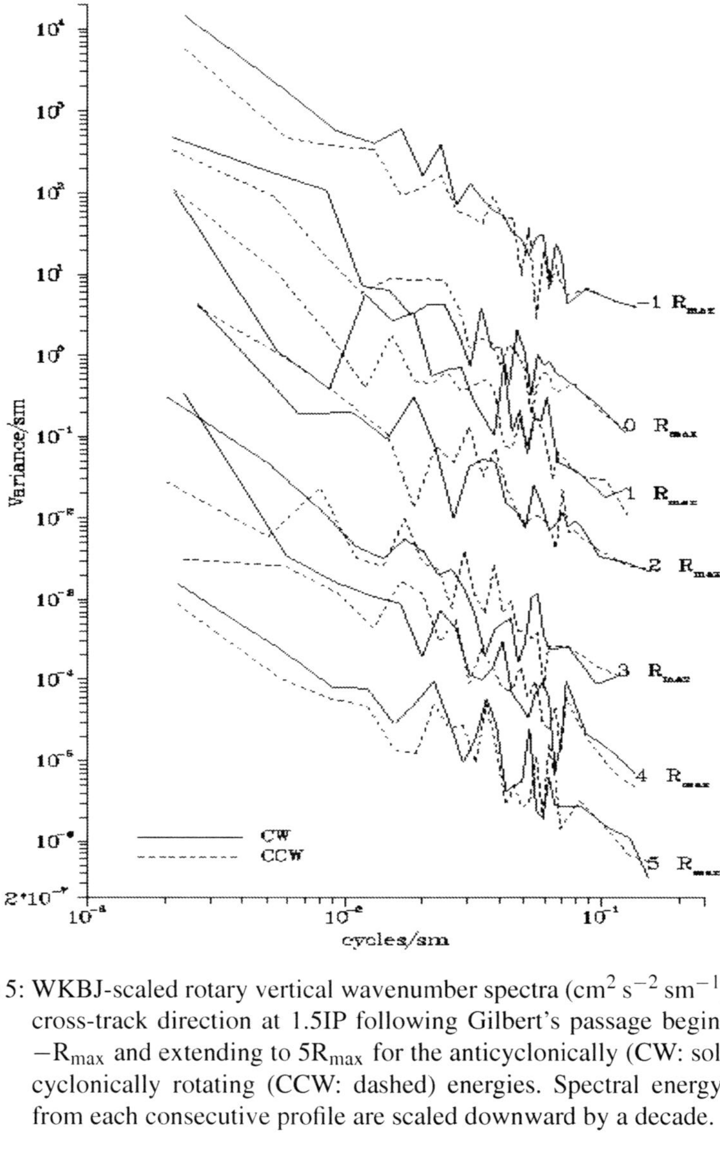


Figure 5: WKBJ-scaled rotary vertical wavenumber spectra ($cm^2\ s^{-2}\ sm^{-1}$) in the cross-track direction at 1.5IP following Gilbert's passage beginning at $-R_{max}$ and extending to $5R_{max}$ for the anticyclonically (CW: solid) and cyclonically rotating (CCW: dashed) energies. Spectral energy levels from each consecutive profile are scaled downward by a decade.

### 3.3 Energy partitioning

WKBJ-scaled rotary spectral energies were integrated over vertical wavenumber space to estimate the partitioning of the CW and CCW energy densities during Gilbert. Of all the velocity profiles, 63 (83%) revealed a predominance of the CW-rotating energy (Fig. 6). Energy ratios from 0 to 1 indicated that 28% of the profiles had energy ratios from 0 to 1, including the 17% where CCW-rotating energy

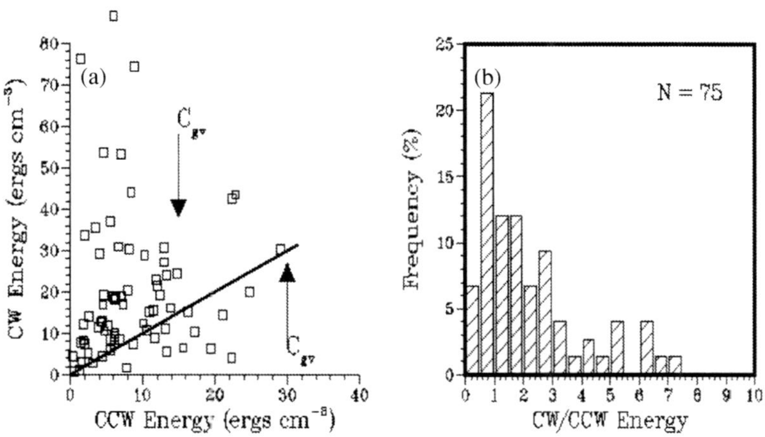


Figure 6: (a) Scatterplot (left panel) of the integrated CW- and CCW-rotating spectral energy densities (ergs cm$^{-3}$), and (b) histogram (right panel) of the ratio of CW($m$) to CCW($m$) spectral energy densities derived from the scaled wavenumber spectra from 75 AXCPs, deployed for Gilbert. The line CW = CCW separates the two counter-rotating regimes. Arrows depict downward (upward) propagating wave group speeds for the CW (CCW) rotating domains, respectively.

was dominant plus the profiles where this ratio was close to unity. The average ratio of the CW- to CCW-rotating components was 3.6, indicative of a preference for downward energy propagation from the wind-forced OML into the thermocline. Correlation levels between the CW and CCW components exceeded 0.74 for the scaled profiles, where the regression slope was about 1.5 and the bias was 15 ergs cm$^{-3}$. A weighted rate of turning of the velocity vector (direction) with the scaled energies was also used to determine the directionality of the downward energy flux from strong forcing events (D'Asaro [29]). Conceptually, both approaches are analogous to a double-helix structure in the vertical with the stronger CW-rotating strand dominating the CCW-rotating strand, corresponding to downward energy propagation and upward phase propagation (Leaman [19]).

### 3.4 Energy fluxes

Excess spectral energy densities were multiplied by the vertical group velocity for each wavenumber contained within the spectra to provide an estimate of the vertical energy flux from the expression

$$F_{\mathrm{iw}} = \int_{2\pi/D}^{2\pi/6} \Delta E(m) c_{\mathrm{gv}}(m)\, \mathrm{d}m, \tag{2}$$

where $D$ represents the profile depth less the start depth of the profiles (nominally 10 m), $\Delta E(m)$ represents the difference between the CW and CCW energies [CW($m$) – CCW($m$)], and $c_{gv}$ is the vertical group velocity. Based on the vertical structure of internal waves with an $e^{imz}$ dependence, Rossby and Sanford [30] determined the vertical group velocity to be

$$c_{gv} \cong -2f\varepsilon m^{-1}, \tag{3}$$

where $\varepsilon$ represents departures from the local Coriolis parameter ($f$) and $m$ is the vertical wavenumber. The departure of the observed frequency from $f$ is ~0.04, as described above. For the largest resolvable vertical wavelength of 2000 m from the profiles, $c_{gv}$ ranged from 0.1 to 0.2 cm s$^{-1}$, which agreed with previous estimates from moored observations (Brooks [31]; Shay and Elsberry [32]) and simulations (Price [23]).

The product of the excess CW energy and vertical group velocity was integrated across the wavenumber spectrum, following eqn (2). For the 63 CW energy-dominated profiles (Fig. 7), downward energy fluxes ranged from 1 to 8 ergs cm$^{-2}$ s$^{-1}$, with an average value of 1.6 ergs cm$^{-2}$ s$^{-1}$. This flux estimate was 5–10 times larger than the flux found by Leaman [19], but was similar to the values of 2 to 4 ergs cm$^{-2}$ s$^{-1}$ obtained in the North Pacific Subtropical Front experiment (Kunze and Sanford [33]). This result suggests that the TC wind field excited as much internal wave energy in the thermocline as that found in the vicinity of oceanic fronts.

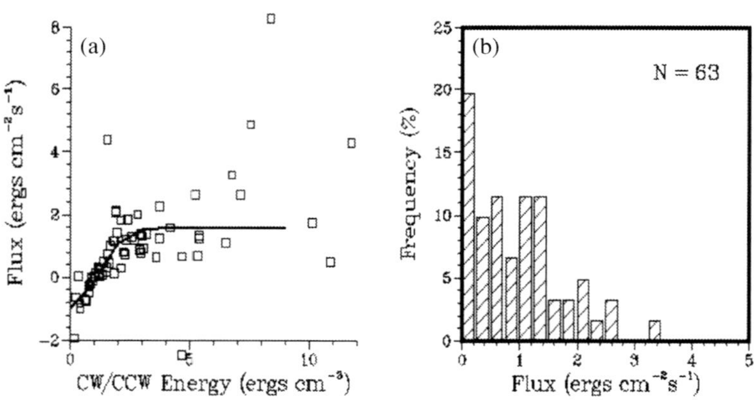


Figure 7: (a) Scatterplot (left panel) of the ER of the anticyclonically (CW) and cyclonically (CCW) rotating spectral energy densities and internal wave energy fluxes (ergs cm$^{-2}$ s$^{-1}$), and (b) histogram (right panel) for the scaled current profiles from 63 AXCPs. The hyperbolic tangent dependence is shown as a solid curve. Notice two distinct regimes in the energy flux scatter, between 0-5 ER and beyond 5 ER, which are accentuated for scaled profiles.

From a global perspective, Nilsson [11] estimated the near-inertial wave power for TCs translating at $7\,\mathrm{m\,s^{-1}}$ to be $\sim 2.6 \times 10^{10}$ W. If a rate of $2\,\mathrm{ergs\,cm^{-2}\,s^{-1}}$ is used as found here, the near-inertial wave power is estimated to be about $7.4 \times 10^{11}$ W, or roughly 30 times this theoretical estimate. Similar estimates of wave power have been found in the conversion of barotropic to internal tides of $3 \times 10^{11}$ W (Bell [34]) and $10^{12}$ W (Sjoberg [35]). The estimated energy flux from TC force winds found here represents a fairly realistic overall value based on these comparisons. Regression analyses between the energy fluxes and the ratio of the scaled energies indicated a hyperbolic tangent relation between these two regimes, of the form:

$$F_{\mathrm{iw}} = 0.2 + 1.4 \tanh\{(\mathrm{CW} \times \mathrm{CCW}^{-1}) - 1.2\}, \tag{4}$$

where a linear relationship exists between the energy flux and the CW/CCW energy ratio (ER). The first regime occurs between ER values of 0–4, and corresponds to an increase in the energy flux for the larger ER values. Beyond ER values of 4.5, a second regime consists of a nearly constant flux region on average, where the hyperbolic tangent function approaches a value of $1.6\,\mathrm{ergs\,cm^{-2}\,s^{-1}}$. The correlation coefficient between energy flux and ER is about 0.7. Note that the ER range from 3 to 4 agrees well with studies on the internal wave spectrum, indicating a preference for CW-rotation of the velocity vector and downward energy propagation for surface-intensified flows.

## 4 Relationship to forcing structure

### 4.1 Approach

Following Kunze and Sanford [33], we assume that the near-inertial energy flux based on eqn (2) is due to surface wind in the following form:

$$F_{\mathrm{iw}} \cong V_{\mathrm{o}} \rho_{\mathrm{a}} c_{\mathrm{d}} U_{10}^{2}, \tag{5}$$

where $V_{\mathrm{o}}$ is the scaled wave group amplitude, $\rho_{\mathrm{a}}$ is the air density, $c_{\mathrm{d}}$ is the surface drag coefficient, and $U_{10}$ is the 10-m wind. Amplitudes of the wave groups were estimated from the near-inertial profiles at the top of the thermocline where the largest energies were found. Scaled wind-driven current amplitudes ranged from 15 to $50\,\mathrm{cm\,s^{-1}}$, using only the profiles for CW-rotating energies associated with downward energy propagation (Fig. 8). The average amplitude of the wave group was $34\,\mathrm{cm\,s^{-1}}$ with a standard deviation of $12\,\mathrm{cm\,s^{-1}}$. By contrast, the background internal wave amplitudes ranged between 3 and $8\,\mathrm{cm\,s^{-1}}$ for the range of $j_*$ values shown in Table 1. Moreover, similar wave group amplitudes of $30\,\mathrm{cm\,s^{-1}}$ were found in moored thermocline current data observed during the passage of TCs Allen (Brooks [31]) and Frederic (Shay and Elsberry [32]), where the response persisted for at least 4 weeks following storm passage. Notice that the scatter between the internal wave energy fluxes and these amplitudes in Fig. 8 indicates a direct relationship. That is, larger energy fluxes were generally associated with

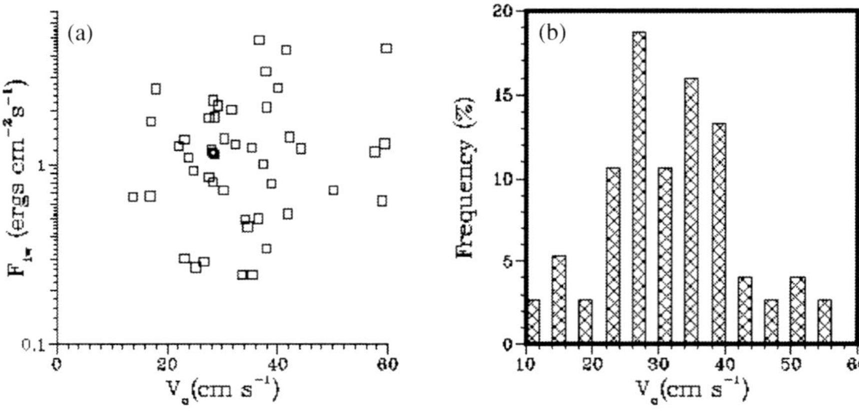


Figure 8: (a) Internal wave flux ($F_{iw}$) versus amplitude of the WKB-scaled wave group ($V_o$). (b) Histogram of $V_o$, for Hurricane Gilbert. $F_{iw}$ is plotted on log scale for clearer depiction of the process and analyses are confined to the region of ~0–5 ER in Fig. 7 for the scaled velocity profiles.

Table 1: Summary of comparisons between the theoretical values from the internal wave spectra ($E_{iw}$), normalized by the WKBJ-scaled observed energies ($E_o$), for a thermocline thickness of 1300 m in the western Gulf of Mexico as a function of $j_*$. The amplitude of the wave group from all of the profilers was ~34 cm s$^{-1}$ with a standard deviation of 12 cm s$^{-1}$.

| $j_*$ | $E_{iw}E_o^{-1}$(%) | $u_{iw}$(cm s$^{-1}$) |
|---|---|---|
| 6 | 23 | 8 |
| 15 | 11 | 4 |
| 20 | 9 | 3 |

larger wave amplitudes, consistent with theoretical considerations. These fluxes and the wave amplitudes were determined from the velocity profiles.

As per eqn (5), these quantities are related to the surface wind stress. Since the winds were observed (as shown in Fig. 3), the air density is taken as $1.16 \times 10^{-3}$ gm cm$^{-3}$. The amplitudes of the wave groups used the average value of 34 cm s$^{-1}$ for fluxes between 0.5 and 2 ergs cm$^{-2}$ s$^{-1}$ and for $V_o$ below and above these fluxes, values of 22 and 46 cm s$^{-1}$, respectively (i.e. 1 standard deviation), were used in this analysis. The remaining unknown is the drag coefficient. The scatter in wind speed squared and the fluxes normalized by the amplitude of the wave group is shown below in Fig. 9. At wind speeds from 10 to 20 m s$^{-1}$, there is a significant increase in the fluxes. The inflection point, or change in the

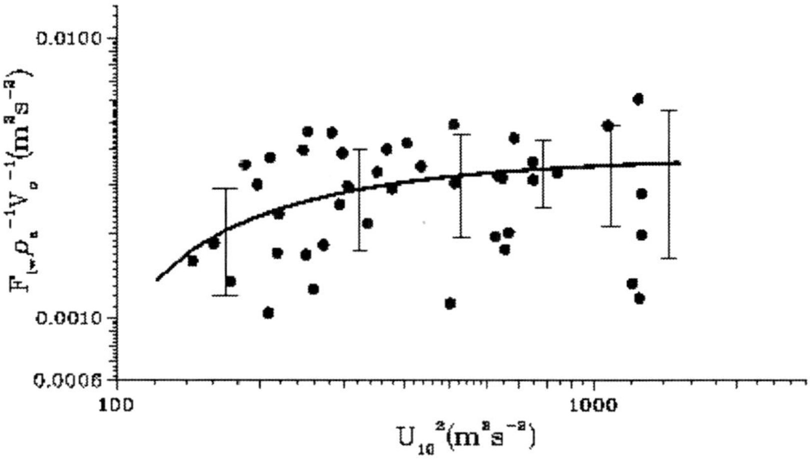


Figure 9: Scatterplot of WKBJ-scaled internal wave flux $F_{\text{iw}}(V_o\rho_a)^{-1}$ versus observed $U_{10}^2$ including the best line fit (solid) and binned standard deviations relative to the regression curve. Analyses are confined for values between $1 \times 10^{-3}$ and $8 \times 10^{-3}$ based on wave fluxes and scaled velocity profiles.

slope occurs at about 28 m s$^{-1}$. This is particularly interesting since recent tank experiments by Alamaro [20] suggest a change in the slope at about this wind speed. Moreover, as winds increased beyond 28 m s$^{-1}$, normalized internal wave fluxes approach a relatively constant value, in an average sense. At very high winds, the scatter becomes much larger in these normalized fluxes. Thus, caution is needed, given the atmospheric variability in the density structure as well as the oceanic variability associated with the LCWCR, where significant exchanges of heat occur because of deep, warm layers (Shay *et al.* [36]).

The fit between the internal wave flux normalized by the air density and the wave group amplitudes from eqn (5) gives a hyperbolic tangent dependence in Fig. 7a

$$\Im(U_{10}) = 3.755 \times 10^{-3} + \tanh(-0.295 \cdot U_{10}^{-2}), \tag{6}$$

valid for observed winds from 10 to 38 m s$^{-1}$. This nonlinear functional dependence suggests that beyond a wind speed of 28 m s$^{-1}$, internal wave fluxes remain relatively constant, in contrast to the large variations occurring for winds between 10 to 20 m s$^{-1}$. This result also suggests that there may be a limit to the amount of energy input into the thermocline ocean current spectrum during TC passage.

### 4.2 Comparisons

From a bulk aerodynamic perspective, Large and Pond [12], the WAMDI group [13] and Smith [16] determined various wind-speed dependent drag coefficients for moderate to strong winds. These drag coefficients ($\times 10^{-3}$) have the following form:

$$c_d = a + b \times U_{10}, \tag{7}$$

where $a$ is the bias and $b$ is the slope of the regression curve, times the $U_{10}$. For Large and Pond [12], $a = 0.49$ and $b = 0.065$, for WAMDI [13] $a = 0.80$ and $b = 0.065$, and for Smith [16] $a = 0.61$ and $b = 0.063$. The Large and Pond [12] formulation is valid over the range from 11 to 25 m s$^{-1}$ whereas the WAMDI [13] formulation has been applied to TCs based on observed and simulated significant wave heights from Hurricanes Camille, Frederic, and Anita. The Smith [16] formulation is valid up to 22 m s$^{-1}$. For a 40 m s$^{-1}$ near-surface wind, drag coefficients range from 3.1–3.5 $\times 10^{-3}$ resulting in a 10–15% uncertainty in the stress, plus errors in $U_{10}$ based on a reduction of the flight-level winds to 10-m (Powell [22]). In many modeling studies, the underlying assumption is that one can extrapolate the drag coefficients to increase ad infinitum—well beyond the range of data used in their formulation. Based on observed wind-driven currents from several TCs, there certainly appears to be a limit in the forced OML currents. Recently Powell *et al.* [21] have suggested a leveling off of $c_d$ at a particular threshold wind speed, whereas Alamaro [20] have suggested that $c_d$ decreases with increased wind speeds beyond a threshold $U_{10}$.

To examine the functional dependence of eqn (6), a comparison is made to the drag coefficient formulations based on wind speed (Large and Pond [12]; the WAMDI Group [13]; Smith [16]) and those based on wave age (Donelan [14]). For example, Alamaro [20] constructed an annular wind-wave flume and used it to estimate surface drag coefficients and the air–sea enthalpy fluxes for equivalent 10-m wind speeds up to 55 m s$^{-1}$. They found that the drag coefficient increases with wind speed but levels off at wind speeds between 26 and 30 m s$^{-1}$, at $c_d \sim 3.8 \times 10^{-3}$. As shown in Fig. 10, this analytical function is essentially the same as that of Large and Pond [12] at low-wind speed. One of the interesting aspects of the hyperbolic tangent dependence of eqn (6) is its rate of increase from 0.002 to 0.003 compared to $c_d$ estimates by Large and Pond [12], in the 10–20 m s$^{-1}$ range. Moreover, there is marked agreement with the wave age-based $c_d$ estimates of Donelan [14], for this $U_{10}$ range. In the present formulation, orbital velocities associated with surface waves have been removed from the profiles following Sanford *et al.* [6], and the residual surface wave-induced currents in the OML have not been included in the analysis as per WKBJ-scaling arguments. In the 20–30 m s$^{-1}$ range, the function approaches a threshold where the slope of the curve approaches 0 at about 28 m s$^{-1}$, and the limit approaches $\approx$0.0035 at higher wind speeds. Up to 25 m s$^{-1}$ in this leveling off region, the Large and Pond [12] $c_d$ increases monotonically with wind speed as does the WAMDI [13] $c_d$, although at a slightly faster rate of increase. At a wind speed of 40 m s$^{-1}$, the value of this analytical function would be nearly equal to that of the WAMDI formulation.

In recent experiments, Global Positioning System (GPS) dropsondes were deployed from research aircraft to measure the quasi-Lagrangian wind profiles

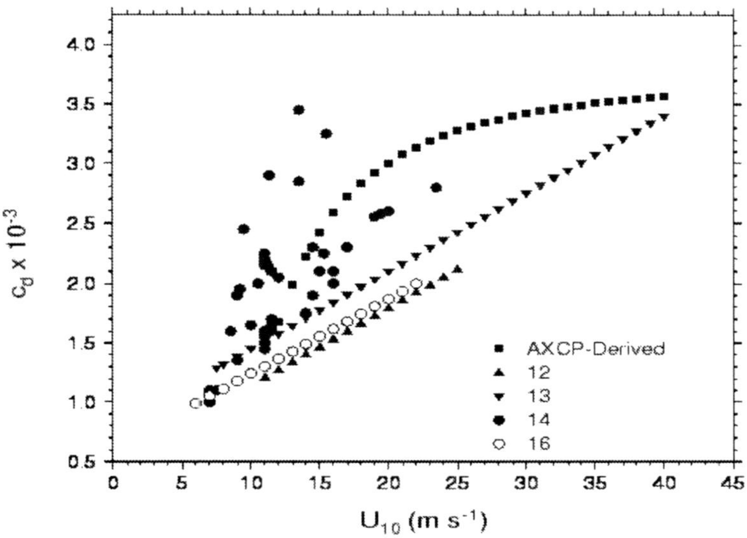


Figure 10: Relationship between the AXCP-derived analytical function (squares) in eqn (6) compared to the wind speed-dependent (Large and Pond [12]; WAMDI [13]; Smith [16]) and wave age-dependent (Donelan [14]) drag coefficients as per the legend.

in the atmospheric boundary layer in TCs. A logarithmic variation of mean wind speed was found in the lowest 200 m, to a maximum of 500 m, with a gradual weakening with height to 3 km (Powell *et al.* [21]). In these estimates, surface stress, roughness length, and neutral stability drag coefficient determined by the profile method also suggest a leveling of the surface momentum flux, and as winds increase above hurricane force, there is a slight decrease of the drag coefficient increasing winds.

There are parallels that can be drawn from the analytical function [eqn (6)] based on the internal wave fluxes and the drag coefficients derived from more conventional approaches. For instance, similar drag coefficient trends were inferred from a model study of storm surges in the North Sea (Mastenbroek *et al.* [37]) as found by a direct evaluation (Amorocho and DeVries [38]) over a range of wind speeds from 0 to 40 m $s^{-1}$. The nonlinear functional dependence suggests that wind stress increase is primarily due to winds beyond 30 m $s^{-1}$. Yet, considerable uncertainty remains at these very high-wind speeds, and caution has to be applied in estimating surface wind stress from any drag coefficients and the analytical function found above [eqn (6)]. Perhaps other factors such as the directionality of the wind, stress, and wave fields need to be accounted for in a hurricane environment as suggested by Geernaert [39]. In a TC environment, however, as the surface wave field is not fully developed, it may lead to more uncertainties in the momentum transfer between the two fluids.

# 5 OML current simulations

Variability in the upper ocean momentum response due to stresses estimated from different drag coefficient formulations is investigated using the primitive HYCOM equation (Bleck [17]). The simulated current structure is compared to the observed current profiles for the four wind stress fields to quantify OML current differences due to these differing formulations.

## 5.1 Model configuration

HYCOM uses a hybrid coordinate system in the vertical that is designed to transition from $z$-levels and isopycnic coordinates in the ocean interior to sigma coordinates in shallow water regions. The model used in this study is implemented for a domain extending from 14° to 31°N and from 80° to 98°W with a horizontal resolution of 0.07°. The model has 250 × 242 horizontal grid points and 50 levels in the vertical. The model is initialized with uniform stratification throughout the domain and therefore the prestorm fields remain quiescent. This contrasts the modeling results of Jacob and Shay [41], where the initial ocean state was not at rest. With a minimum vertical spacing of 3 m, the upper 30 model levels remain in the $z$-coordinate system. The model is configured with the K-profile vertical mixing parameterization of Large *et al.* [40]. The boundaries along the Florida Straits and the Caribbean Sea are closed by vertical sidewalls. The model is integrated for 6 days using the stresses derived from realistic wind fields covering Storm, Wake 1, and Wake 2 observational snapshots (Fig. 2).

Simulated OML currents are defined as a vertical integral from the surface to the depth of the layer. The OML depth is defined as the level where the temperature changes by more than 0.2°C, which is consistent with the thermistors on the AXCPs (Shay *et al.* [5]). While Jacob and Shay [41] showed that determining accurate OML depths from an isopycnic model is difficult, the HYCOM model circumvents this issue with $z$-coordinates in the OML. This provides a closer depiction of the observed currents and densities within the layer.

## 5.2 Surface stresses

Bulk drag coefficient formulations from Large and Pond [12], WAMDI [13], Smith [16], and the AXCP-derived function [eqn (6)] are used to estimate wind stresses for forcing the ocean model. With a maximum wind speed of $\sim$38 m s$^{-1}$ associated with Gilbert, the drag coefficients are extrapolated as necessary. For the AXCP-derived function [eqn (6)], a constant coefficient of $1.14 \times 10^{-3}$ is used for wind speeds less than 10 m s$^{-1}$. Estimated stresses for winds in Fig. 3 suggest a maximum difference of 20% for the four drag coefficient formulations, as shown in Fig. 11, with Smith [16] and the AXCP-derived function giving the minimum and maximum stresses, respectively.

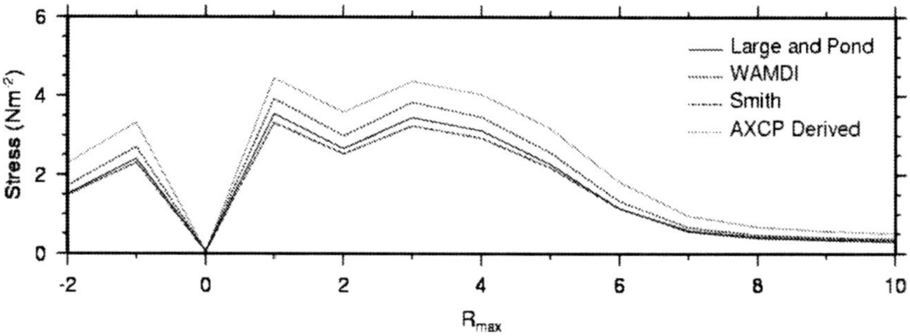


Figure 11: Cross-section ($R_{\max}$) of estimated surface wind stresses ($\mathrm{N\,m^{-2}}$) from Large and Pond [12], WAMDI [13], Smith [16] $c_{\mathrm{d}}$ drag coefficient parameterizations and that inferred from the AXCP-derived analytical function [eqn (6)] estimated using the wind field in Fig. 3.

### 5.3 Comparisons

As the internal wave fluxes were derived for scaled current profiles, observed OML currents were not included in the analysis ($N = 0$ in the OML). Thus, to examine the sensitivity of the derived analytical function [eqn (6)], simulated OML currents are compared to those observed for the storm snapshot (Fig. 2) of Hurricane Gilbert as well as the other snapshots. As shown in Fig. 12, the horizontal structure of the simulated current response in the OML during the storm snapshot suggests good agreement among the four cases. Maximum simulated mixed layer currents occur for the AXCP-derived function [eqn (6)], corresponding to maximum stress values. In this case, stronger OML currents exist away from the track due to higher stress values, compared to the other three formulations. However, the overall OML currents are slightly weaker in the AXCP-derived model compared to observations. While the near-surface currents in this model are stronger, due to the rapid decay of currents with depth, averaging within the OML reduces this current magnitude. Maximum values are within ~10–15% of the observed currents for the AXCP-derived formulation.

Simulated mixed layer currents for the four drag coefficients are compared to the observed currents from AXCP data using linear regression statistics for Hurricane Gilbert, including the total data set of all three snapshots in Tables 2 and 3. While RMS differences between the drag coefficient formulations remain quantitatively similar, the slope of the regression line suggests a marginally better fit for the AXCP-derived function [eqn (6)] to the current data, as shown in Fig. 13. Compared to previous layered-model simulations of currents by Jacob and Shay [41], the magnitude of simulated OML currents from HYCOM are closer to observed values. For the Smith [16] drag coefficient (not shown) similar agreements were found. Using data from all three snapshots did not significantly reduce the RMS differences.

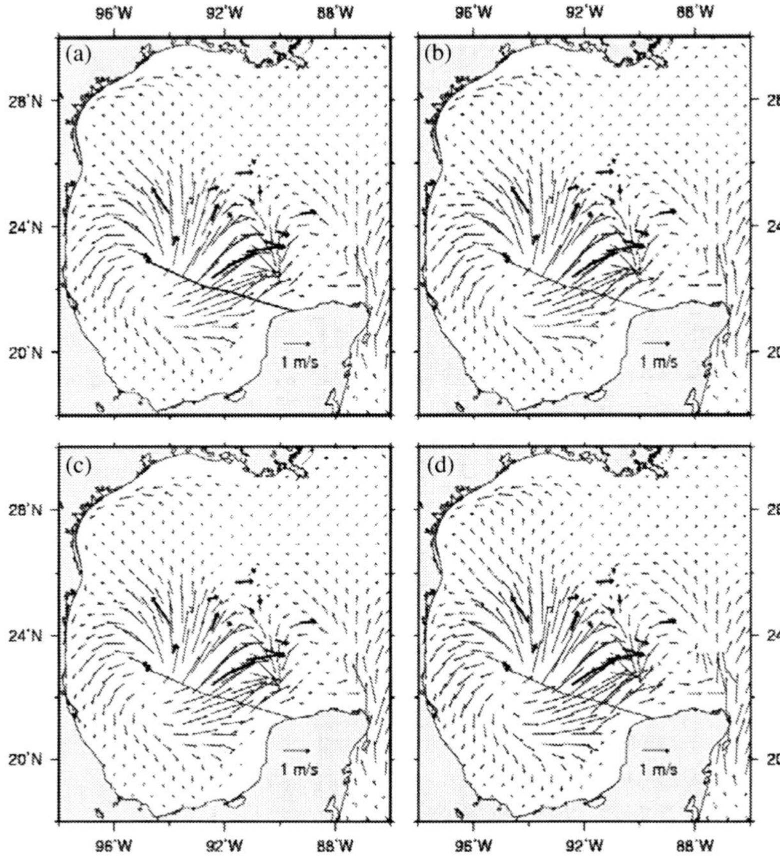


Figure 12: Simulated OML currents for the four stress distributions in Figure 11 for a quiescent ocean for (a) Large and Pond [12], (b) WAMDI [13], (c) Smith [16] and (d) AXCP-derived function (eqn 6) where thicker arrows are observed storm-induced currents. For the AXCP-derived function (d), currents are stronger extending farther away from the storm track (solid black line).

A key question emerging from our analysis concerns the qualitative similarities between the simulated currents implicit in an accepted drag coefficient such as WAMDI [13] compared to those based on the AXCP-derived analytical function [eqn (6)]. As shown in Fig. 14, the comparisons suggest a direct relationship for both current components. Both simulated current components are well predicted by the analytical function [eqn (6)] compared to the WAMDI [13] drag formulation, with a spread of $\sim 10\,\mathrm{cm\,s^{-1}}$, relative to the ideal line with a slope of unity. This is the noise level of AXCP observations in a hurricane environment (Sanford *et al.* [6]). These results suggest a tight relationship. Given this level of agreement between

Table 2: Linear regression statistics quantifying the differences between observed storm currents and simulated currents for four different drag coefficients. All values are expressed as $cm\,s^{-1}$ except the regression curve slope.

| | Large and Pond [12] | | WAMDI Group [13] | | Smith [16] | | tanh | |
|---|---|---|---|---|---|---|---|---|
| $U$: Slope/bias | 0.78 | 7 | 0.82 | 7 | 0.76 | 8 | 0.83 | 9 |
| $V$: Slope/bias | 1.13 | 17 | 1.17 | 19 | 0.73 | 17 | 1.14 | 23 |
| Mean $U$ | 0 | | −1 | | 1 | | −3 | |
| Mean $V$ | −19 | | −22 | | −18 | | −25 | |
| $\sigma_u$ | 48 | | 50 | | 48 | | 54 | |
| $\sigma_v$ | 31 | | 33 | | 30 | | 35 | |
| RMS $U$ | 47 | | 49 | | 47 | | 53 | |
| RMS $V$ | 36 | | 39 | | 34 | | 43 | |

Table 3: As in Table 2, except all three snapshots (defined in Fig. 2) of mixed layer currents for four different drag coefficient parameterizations.

| | Large and Pond [12] | | WAMDI Group [13] | | Smith [16] | | tanh | |
|---|---|---|---|---|---|---|---|---|
| $U$: Slope/bias | 0.88 | 7 | 0.91 | 8 | 0.85 | 8 | 0.89 | 8 |
| $V$: Slope/bias | 1.05 | 10 | 1.08 | 13 | 1.00 | 0.11 | 1.10 | 14 |
| Mean $U$ | −5 | | −7 | | −5 | | 8 | |
| Mean $V$ | −11 | | −13 | | −11 | | −14 | |
| $\sigma_u$ | 44 | | 46 | | 42 | | 46 | |
| $\sigma_v$ | 34 | | 35 | | 33 | | 38 | |
| RMS $U$ | 44 | | 45 | | 43 | | 47 | |
| RMS $V$ | 36 | | 38 | | 34 | | 41 | |

WAMDI [13] and the tanh function [eqn (6)], the results suggest that perhaps this latter function may reasonably represent the drag coefficient: simulated currents are certainly as good as those resulting from the extrapolated Large and Pond [12] and Smith [16] values, and are consistent with results from the WAMDI [13] drag coefficient.

While the accepted drag coefficients continue to increase the surface wind stress, the analytical function [eqn (6)] and drag coefficients suggested by Alamaro [20] and Powell *et al.* [21] tend to level off at a particular threshold wind speed as mentioned in Section 4.2. The black line is for all profiles from three snapshots (see Fig. 3). Bias and slopes of the regression curves for the storm snapshot (only) are given in the lower left of each panel.

However, the major differences between the drag coefficients of Large and Pond [12], WAMDI [13], and this AXCP-derived function [eqn (6)] may only become

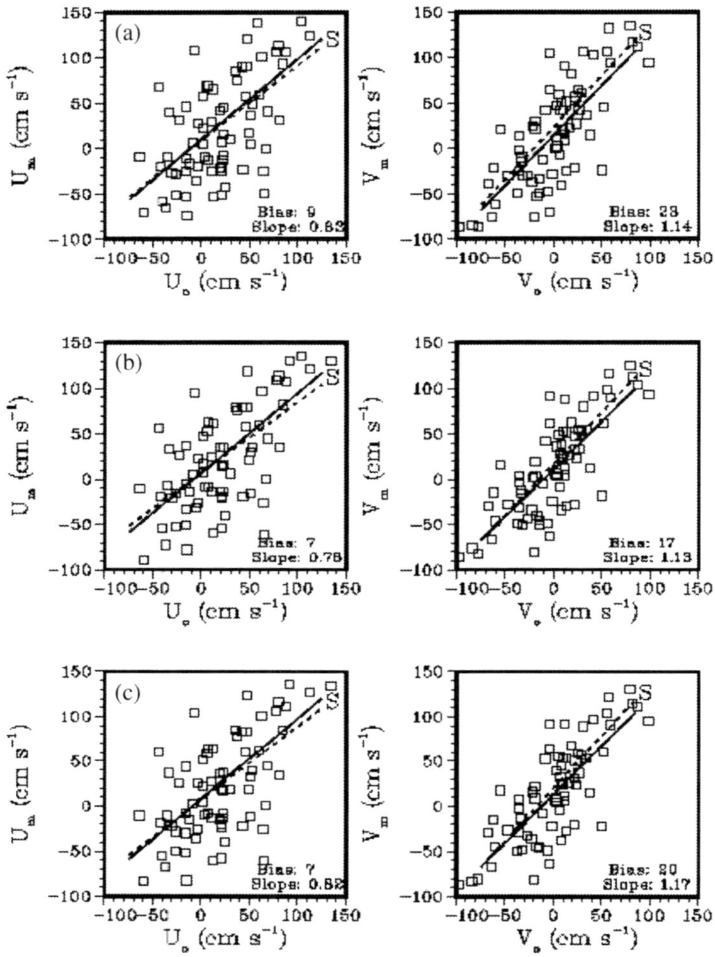


Figure 13: Comparison of simulated and observed OML currents for the $U$-component (left panels) and $V$-component (right panels) for (a) AXCP-derived analytical function, (b) Large and Pond [12], and (c) WAMDI [13] drag coefficients. The dashed line is for least squares regression fit, for Hurricane Gilbert, snapshot only.

more apparent for winds exceeding $40\,\mathrm{m\,s^{-1}}$, in terms of dynamical processes required to keep the stress nearly constant or to decrease for winds above this threshold. This issue will be further investigated using upper ocean current measurements from AXCPs acquired during higher surface winds such as those recently observed in TCs Isidore and Lili as they deepened to category 3 and 4 storms. Notwithstanding, the importance of acquiring high-resolution current profiles during strong wind events is very useful in understanding the amount of momentum being input to the OML by the wind stress (Sanford *et al.* [6]). Our results also point to the fact that

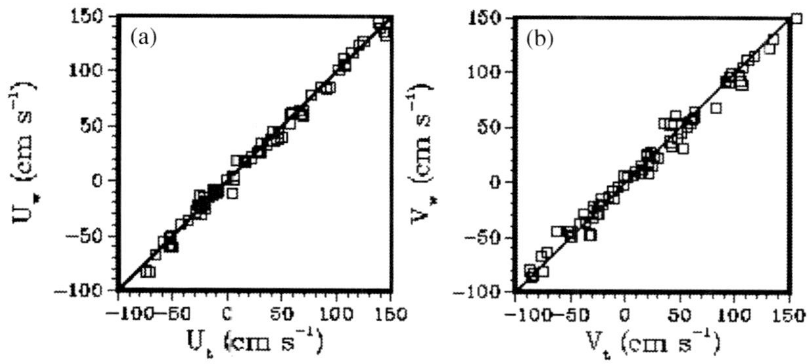


Figure 14: Scatterplots between simulated currents from AXCP-derived tanh [eqn (6)] function (abscissa) and the WAMDI drag coefficient (ordinate) for (a) $U$-component and (b) $V$-component of current (cm s$^{-1}$). The solid black line represents the unity slope line.

the ocean can be used as a tracer of the momentum flux, when high-resolution air–sea flux measurements may not be possible under very strong wind and wave conditions.

## 6 Concluding remarks

There has been significant progress in improving our understanding of the basic oceanic and atmospheric processes that occur during the passage of TCs. Central to this progress is the need to isolate fundamental physical processes involved in the air–sea interactions through detailed process studies using experimental, empirical, theoretical, and numerical approaches. As has been found from previous sets of measurements (Hurricane Gilbert, Spectrum 90, Pudov and Petrichenko [42]), these approaches are needed to eventually improve predictions of storm intensity and structure (Shay [43]).

The goal of the Gilbert experiment in September 1988 was to measure the evolution of the three-dimensional current and temperature structure excited by the passage of a TC over the scales of the atmospheric forcing. For the first time, the evolving three-dimensional oceanic response to the passage of a TC was accurately determined from the successful deployment of AXCPs in the western Gulf of Mexico (Shay [1]). The sampling strategy resulted in a reliable time–space series to address the three-dimensional evolution of the near-inertial wave wake (Shay *et al.* [9]) and the heat and mass balance (Jacob *et al.* [24], Jacob and Shay [41]). This approach has recently been duplicated and enhanced during the passage of TCs Isidore and Lili during deepening cycles in the Gulf of Mexico and Caribbean Sea in 2002.

Current profiles revealed a CW-rotation of the current vector with depth, indicating vertical propagation of near-inertial wave groups from the OML into the thermocline. CW-rotating spectral energies dominated the scaled energy spectra in the profiles by an average factor of four, and followed a tanh dependence in the ratio of CW and CCW energies. Downward energy fluxes ranged from 0.5 to 10 ergs $cm^{-2}$ $s^{-1}$ with an average value of ~2 ergs $cm^{-2}$ $s^{-1}$ for scaled profiles. These energy fluxes were 4–10 times larger than in unforced cases (Leaman and Sanford [25]), but were similar to those found in frontal regimes (Kunze and Sanford [33]).

By comparing the internal wave fluxes to the square of the wind speed, an analytical function was found that suggests a nonlinear dependence on the wind speed. This approach differs considerably from previous treatments of the wind stress by using the oceanic response as a tracer of the amount of momentum flux transferred to the oceanic thermocline from the forced OML, and is consistent with the propagation of energy arguments. As shown in previous studies, we found that the response is dominated by near-inertial wave processes where the wave group amplitudes are observed at the top of the thermocline (Shay *et al.* [5]; Brooks [31]; Shay and Elsberry [32]). Moreover, as the vertical wavenumber spectral energy densities are between 4–10 times larger than the background internal wave spectrum (Smith [16]; Garrett and Munk [18]), therefore, the wave power was found to be at least 30 times larger than theoretical estimates from a global perspective (Nilsson [11]). It is no surprise that hurricane forcing of the thermocline current response, which may persist for at least a month following storm passage, is as effective as oceanic fronts in focusing internal wave energy close to the near-inertial frequency ($\sim f$).

The derived analytical function, based upon three sets of point measurements, agrees well with the Large and Pond [12] formulation at a wind speed of 11 m $s^{-1}$, and at about 40 m $s^{-1}$ for the WAMDI [13] drag coefficient. There is fairly good agreement between these two wind regimes and the wave age-dependent drag coefficient of Donelan [14]. The latter suggests a considerably higher drag coefficient between 10 and 20 m $s^{-1}$, as reflected by the analytical function. At a threshold wind speed of ~28 m $s^{-1}$, the slope of the function approaches 0 suggesting that beyond this wind speed the internal wave spectrum tends to saturate as in the case of surface waves. Comparison of this analytical function to other accepted drag coefficients indicates that there may also be a limit to the wind-speed dependence of the drag coefficient. Beyond about 28 m $s^{-1}$, near-surface wind stress may be dependent solely on wind speed increases, rather than monotonically increasing functions.

Detailed comparisons using these formulations in HYCOM were completed. Since the AXCP-derived function is based on energy entering the thermocline (profiles were WKBJ-scaled), simulated OML currents compared to observed currents represent an independent test. While RMS differences in simulated and observed currents were quite large for all cases, the analytical AXCP-derived function suggested currents did as well as those implicit in extrapolated (Large and Pond [12]; WAMDI [13]; Smith [16]) drag coefficients. Results from the WAMDI [13] formulation were based on buoy data from three hurricanes. The regression slope for the function is also slightly larger compared to the other methods relative to the observed OML currents. The results suggest that this issue needs to be looked at from both indirect approaches (as described here) and direct measurements.

Finally, there is a relationship between surface winds and the internal wave flux, which is dominated by near-inertial processes. While the analytical function is not a conventional drag coefficient in the strictest sense (as determined by the more direct and higher-resolution measurements), using the oceanic current response as a tracer of input by the wind stress not only opens new questions concerning the momentum flux at high wind speeds but also raises the possibility of using oceanic current measurements as a tracer in understanding the momentum transfer between the two fluids (Smith *et al.* [15]). Clearly, this issue needs further inquiry using high-resolution measurements of the surface wind and wave processes with concurrent oceanic velocity and atmospheric wind profiles. Given uncertainties in ocean surface variables during strong winds, the air–sea exchange coefficients need to be better understood and parameterized for oceanic and coupled models to realistically simulate the oceanic response and feedback to strong forcing events.

## Acknowledgments

L.K.S. gratefully acknowledges support from NSF (ATM-9714885, 0108218) and NOAA. S.D.J. is supported by NASA and NOAA. Mark Powell and Sam Houston (NOAA-HRD) kindly facilitated the use of their software in deriving the wind analyses, and Kevin Leaman kindly shared insights regarding the theme of this paper. Gregg Holland and Mark Donelan provided guidance and stimulating ideas. George Halliwell provided guidance in setting up HYCOM for these hurricane simulations. The TMI SST data in Fig. 1 were provided by Chelle Gentemann from Remote Sensing Systems. Tom Cook kindly provided programming support and Lizett Bowen provided much needed assistance in formatting the document.

## References

[1] Shay, L.K., Upper ocean structure: response to strong forcing events. *Encyclopedia of Ocean Sciences*, eds R.A. Weller, S.A. Thorpe & J. Steele, Academic Press International: London, pp. 3100–3114, 2001.

[2] Geisler, J.E., Linear theory on the response of a two-layer ocean to a moving hurricane. *Geophysical Fluid Dynamics*, **1**, pp. 249–272, 1970.

[3] Gill, A.E., On the behavior of internal waves in the wakes of storms. *Journal of Physical Oceanography*, **14**, pp. 1129–1151, 1984.

[4] Kundu, P. & Thomson, R.E., Inertial oscillations due to a moving front. *Journal of Physical Oceanography*, **15**, pp. 1076–1084, 1985.

[5] Shay, L.K., Elsberry, R.L. & Black, P.G., Vertical structure of the ocean current response to a hurricane. *Journal of Physical Oceanography*, **19**, pp. 649–669, 1989.

[6] Sanford, T.B., Black, P.G., Haustein, J., Fenney, J.W., Forristall, G.Z. & Price, J.F., Ocean response to hurricanes. Part I: Observations *Journal of Physical Oceanography*, **17**, pp. 2065–2083, 1987.

[7] Price, J.F., Sanford, T.B. & Forristall, G.Z., Observations and simulations of the forced response to moving hurricanes, *Journal of Physical Oceanography*, **24**, pp. 233–260, 1994.

[8] Shay, L.K., Black, P.G., Mariano, A.J., Hawkins, J.D. & Elsberry, R.L., Upper ocean response to hurricane Gilbert. *Journal of Geophysical Research*, **97**, pp. 20227–20248, 1992.

[9] Shay, L.K., Mariano, A.J., Jacob, S.D. & Ryan, E.H., Mean and near-inertial ocean current response to hurricane Gilbert. *Journal of Physical Oceanography*, **28**, pp. 858–889, 1998.

[10] Zervakis, V. & Levine, M.D., Near-inertial energy propagation from the mixed layer: theoretical considerations. *Journal of Physical Oceanography*, **25**, pp. 2872–2889, 1995.

[11] Nilsson, J., Energy flux from traveling hurricanes to the oceanic internal wave field. *Journal of Physical Oceanography*, **25**, pp. 558–573, 1995.

[12] Large, W.G. & Pond, S., Open ocean momentum flux measurements in moderate to strong wind. *Journal of Physical Oceanography*, **11**, pp. 324–336, 1981.

[13] WAMDI Group, The WAM model—a third generation ocean wave prediction model. *Journal of Physical Oceanography*, **18**, pp. 1775–1810, 1988.

[14] Donelan, M., Air-sea interaction. *The Sea: Ocean Engineering Series*, 9, John Wiley and Sons Publishing Company: New York, pp. 239–292, 1990.

[15] Smith, S.D., Fairall, C., Geernaert, G.L. & Hasse, L., Air-sea fluxes: 25 years of progress. *Boundary-Layer Meteorology*, **78**, pp. 247–290, 1996.

[16] Smith, S.D., Wind stress and heat flux over the ocean in gale force winds. *Journal of Physical Oceanography*, **10**, pp. 709–726, 1980.

[17] Bleck, R., An oceanic general circulation model framed in hybrid isopycnic-Cartesian coordinates. *Ocean Modeling*, **4**, pp. 55–88, 2002.

[18] Garrett, C.J.R. & Munk, W., Space-time scales of internal waves—a progress report. *Journal of Geophysical Research*, **80**, pp. 291–297, 1975.

[19] Leaman, K.D., Observations of vertical polarization and energy flux of near-inertial waves. *Journal of Physical Oceanography*, **6**, pp. 894–908, 1976.

[20] Alamaro, M., Wind wave tank for experimental investigation of momentum and enthalpy transfer from the ocean surface at high wind speed. M.S. Thesis, MIT Department of Earth, Atmospheric and Planetary Sciences, Cambridge, MA, 77 pp., 2001.

[21] Powell, M., Vickery, P.J. & Reinhold, T.A., Boundary layer wind profiles and drag coefficient behavior in tropical cyclones. *Nature*, **422**, pp. 279–283, 2003.

[22] Powell, M.D., Evaluations of diagnostic boundary layer models to hurricanes. *Monthly Weather Review*, **115**, pp. 2479–2506, 1980.

[23] Price, J.F., Internal wave wake of a moving storm. Part I: Scales, energy, budget and observations. *Journal of Physical Oceanography*, **13**, pp. 949–965, 1983.

[24] Jacob, S.D., Shay, L.K., Mariano, A.J. & Black, P.G., The three-dimensional mixed layer heat balance during hurricane Gilbert. *Journal of Physical Oceanography*, **30**, pp. 1407–1429, 2000.

[25] Leaman, K.D. & Sanford, T.B., Vertical energy propagation of inertial waves: a vector spectral analysis of velocity profiles. *Journal of Geophysical Research*, **80**, pp. 1975–1978, 1975.

[26] Munk, W., Internal waves and small-scale processes. *Evolution of Physical Oceanography: Scientific Surveys in Honor of Henry Stommel*, ed. C. Wunsch, The MIT Press: Massachusetts, pp. 264–291, 1981.

[27] Gonella, J., A rotary component method for analyzing meteorological and oceanographic time series. *Deep-Sea Research*, **19**, pp. 833–846, 1972.

[28] Otnes, R.K. & Enochson, L., *Applied Time Series Analysis*. John Wiley and Sons: New York, 1978.

[29] D'Asaro, E.A., Energy flux from the wind to near-inertial motions in the surface mixed layer. *Journal of Physical Oceanography*, **8**, pp. 1043–1059, 1985.

[30] Rossby, H.T. & Sanford, T.B., A study of velocity profiles through the thermocline. *Journal of Physical Oceanography*, **6**, pp. 766–774, 1976.

[31] Brooks, D.A., The wake of hurricane Allen. *Journal of Physical Oceanography*, **13**, pp. 117–129, 1983.

[32] Shay, L.K. & Elsberry, R.L., Near-inertial ocean current response to hurricane Frederic. *Journal of Physical Oceanography*, **17**, pp. 1249–1269, 1987.

[33] Kunze, E. & Sanford, T.B., Observations of inertial waves in a front. *Journal of Physical Oceanography*, **14**, pp. 566–581, 1984.

[34] Bell, T.H. Jr., Topographically generated internal waves in the open ocean. *Journal of Geophysical Research*, **80**, pp. 320–327, 1975.

[35] Sjoberg, B. & Stigebrandt, A., Computations of the geographical distribution of the energy flux to mixing processes via internal tides and the associated vertical circulation in the ocean. *Deep-Sea Research*, **39**, pp. 269–291, 1992.

[36] Shay, L.K., Goni. G.J. & Black, P.G., Effects of a warm ocean feature on Hurricane Opal. *Monthly Weather Review*, **128**, pp. 1366–1383, 2000.

[37] Mastenbroek, C., Burgers, G. & Janssen, P.A.E., The dynamical coupling of a wave model and a storm surge model. *Journal of Physical Oceanography*, **23**, pp. 1856–1866, 1993.

[38] Amorocho, J. & DeVries, J.J., A new evaluation of the wind stress coefficient over water surfaces. *Journal of Geophysical Research*, **85**, pp. 433–442, 1980.

[39] Geernaert, G., Measurement of the angle between the wind and the wind stress vector in the surface layer over the North Sea. *Journal of Geophysical Research*, **93**, pp. 8215–8220, 1988.

[40] Large, W.G., McWilliams, J.C. & Doney, S.C., Oceanic vertical mixing: a review and a model with nonlocal boundary layer parameterization. *Reviews of Geophysics*, **32**, pp. 363–403, 1994.

[41] Jacob, L.K. & Shay, L.K., The role of oceanic mesoscale features on the tropical cyclone-induced response. *Journal of Physical Oceanography*, **33**, pp. 649–676, 2003.
[42] Pudov, V. & Petrichenko, S., Trail of a typhoon in the salinity field of the upper ocean layer. *Izvestiya, Atmospheric and Oceanic Physics*, **36**, pp. 645–650, 2000.
[43] Shay, L.K., Air-ocean interface and boundary processes. *Rapporteurs Report for Tropical Cyclone Structure and Structure Change*, World Meteorological Organization Tropical Meteorology Research Series, ed. Russell L. Elsberry, WMO Report 67, December 3–12, 2002, Cairns, Australia, 2002.

# CHAPTER 6

# Midlatitude storm impacts on air–sea $CO_2$ fluxes

W. Perrie[1,2], W. Zhang[1,2], X. Ren[3], Z. Long[1,2] & J. Hare[4]
[1] *Fisheries & Oceans Canada, Bedford Institute of Oceanography, Dartmouth, Canada.*
[2] *Department of Engineering Mathematics, Dalhousie University, Halifax, Canada.*
[3] *Department of Atmospheric Sciences, Nanjing University, Nanjing, China.*
[4] *CIRES, University of Colorado and NOAA ETL, Boulder, Colorado, USA.*

## Abstract

The rate of air–sea $CO_2$ exchange is significantly influenced by marine storms because storm-related processes change $fCO_{2w}$, the fugacity of $CO_2$ in bulk water. These processes include storm-induced sea surface temperature cooling, through entrainment mixing at the bottom of the mixed layer and upwelling, and sea surface heat and mass fluxes. Associated processes are sea spray and wave-induced surface roughness. To study these processes, we numerically simulated extratropical Hurricane Gustav (2002), using a coupled atmosphere–ocean wave-spray model. Four recent formulations for the gas transfer velocity $k_L$ and $CO_2$ flux are compared.

## 1 Introduction

Although the effect of hurricanes on the thermal and physical structure of the upper ocean is comparatively well known, the mechanism behind their impact on air–sea $CO_2$ fluxes has not been fully determined. There are few data sets documenting the impacts of hurricanes on air–sea $CO_2$ transfer processes. A good example is a recent study by Bates *et al.* [1] who reported that a 60 μatm $fCO_{2w}$ decrease occurred after Hurricane Felix (1995) passed near Bermuda, with concomitant upwelling and 4°C sea surface temperature (SST) cooling. These processes can modify air–sea $CO_2$ fluxes. Many studies have reviewed the impact of decreased air–sea heat exchanges associated with SST cooling on storm intensity (Ren *et al.* [2]; Chan *et al.* [3]; Jacob *et al.* [4]; Bender and Ginis [5]; Schade and Emanuel [6]). In tropical cyclones, the coupling between the storm and the underlying ocean can result in SST cooling by as much as 6–8°C, and near-inertial currents as large as 2–3 $\mathrm{m\,s^{-1}}$.

Here, we present simulations of Hurricane Gustav as a case study of the relation of these processes to $CO_2$ transfers (Perrie *et al.* [7]). Gustav was an extratropical transitioning storm in the Northwest Atlantic in September 2002. It is selected because measurements of $fCO_{2w}$ and $fCO_{2a}$ (fugacity of $CO_2$ in bulk water air, respectively) were collected by the NOAA Ship Ronald H. Brown (hereafter "Ron Brown") near Gustav's storm track, for the Ocean-Atmosphere Carbon Exchange Study (OACES, http://www.aoml.noaa.gov/ocd/oaces/). A description of the coupled model system is given in Section 2, and a discussion of our simulations of Gustav is given in Section 3. Four recent gas transfer velocity $k_L$ parameterizations are given in Section 4: Wanninkhof [8], Wanninkhof and McGillis [9], Zhao *et al.* [10], and Fairall *et al.* [11]. Estimates of air–sea fluxes are given in Section 5.

## 2 Model description

Our understanding of the impact of Gustav on the air–sea exchange of $CO_2$ depends on a good storm simulation. Our coupled ocean–atmosphere dynamical model system is described by Ren *et al.* [2] and Zhang *et al.* [12]. The model consists of the MC2 (Mesoscale Compressible Community) atmospheric model, the Princeton Ocean Model (POM), a sea spray parameterization, and a wave model. The MC2–POM–WW3-spray model is hereafter denoted the "coupled model".

MC2 is a state-of-the-art, fully elastic, nonhydrostatic model solving the full Euler equations on a limited-area Cartesian domain (Benoit *et al.* [13]). Lateral boundary and initial conditions are taken from the Canadian Meteorological Centre (CMC) analysis data. The model domain is (79.5°W–40.0°W, 24.25°N–56.25°N), using a 0.25° resolution latitude–longitude projection, 30 vertical layers, and 600 s time steps. Over the sea, MC2's interfacial momentum and heat fluxes assume the Monin–Obukhov theory, with a bulk turbulent flux formulation and turbulent transfer coefficients. The latter depend on empirical similarity functions $\psi_m$ and $\psi_h$ and roughness lengths for wind speed, temperature, and humidity, $z_{0m}$, $z_{0t}$, and $z_{0q}$. The bulk spray flux algorithm is based on the analysis of HEXOS data by Andreas [14] and Andreas and DeCosmo [15], and is tuned to the COARE-type (Fairall *et al.* [16]) turbulent bulk flux code that he applied. Thus, we replace MC2's default $z_{0m}$, $z_{0t}$, $z_{0q}$, $\psi_m$, $\psi_h$ relations with the COARE-type relations used by Andreas' [14] spray analysis. Total momentum $\tau_T$, latent $H_{L,T}$, and sensible $H_{s,T}$ fluxes at the lowest model level are obtained by adding the bulk interfacial ($\tau, H_L, H_s$) and spray fluxes ($\tau_{sp}, Q_{L,sp}, Q_{S,sp}$),

$$\tau_T = \tau - \tau_{sp}, \tag{1}$$

$$H_{\mathrm{L,T}} = H_{\mathrm{L}} - Q_{\mathrm{L,sp}}, \quad (2)$$

$$H_{\mathrm{s,T}} = H_{\mathrm{s}} - Q_{\mathrm{S,sp}}. \quad (3)$$

POM (Mellor [17]) is implemented on a 0.16° resolution latitude–longitude projection for (82°W–40°W, 20°N–57.5°N), with 23 layers. Initial and boundary conditions for temperature and salinity are monthly Generalized Digital Environmental Model data (Bender and Ginis [5]), and lateral prescribed barotropic transports. The model spin-up involves integration for 1 year, using monthly mean wind stress, heat flux, and fresh water flux NCEP (National Centers for Environmental Prediction) data. To realistically simulate the prestorm ocean, the integration is continued for a second year using NCEP data for a given storm.

The wave model is WAVEWATCH III (Tolman and Chalikov [18]; Tolman [19]), hereafter denoted WW3, implemented at 0.25° resolution on the same domain as MC2. WW3 gives the directional wave spectra by solving the well-known wave spectral action balance equation in terms of: wind input energy to waves ($S_{\mathrm{in}}$), wave-dissipation ($S_{\mathrm{ds}}$), and wave–wave interactions ($S_{\mathrm{nl}}$). All simulations assume WAM cycle 3 formulations for $S_{\mathrm{in}}$ and $S_{\mathrm{ds}}$.

Model exchanges between the atmosphere and ocean occur at each coupling time step—when MC2, POM, and WW3 time steps are coincident. Wind stress, sensible and latent heat fluxes, radiative flux, and fresh water flux, from MC2 (including spray), are transferred to POM and WW3. New POM-produced SST fields are passed to MC2. Storm-induced SST cooling reduces the sea surface heat fluxes, which leads to reduced storm intensity. This is in competition to spray, which tends to enhance air–sea heat fluxes, and storm intensity (Andreas and Emanuel [20]), whereas enhanced wave drag, resulting from young rough waves, tends to reduce storm intensity.

## 3 Case study: extratropical Hurricane Gustav

Gustav was designated a tropical storm by 12 UTC on September 10, 2002, north of the Bahamas. Nearing Cape Hatteras, Gustav turned northeastward and accelerated, embedded within a southwesterly flow resulting from a baroclinic cyclogenesis over New England. It began to intensify over 28°C Gulf Stream waters and merge with the nontropical low, dominated by baroclinic processes. Gustav became a hurricane near 12 UTC on September 11, reaching maximum intensity of 85 kt (43 m s$^{-1}$) near 18 UTC on September 11. It made landfall over Cape Breton with 80-kt winds near 06 UTC on September 12. Gustav transitioned to an extratropical cyclone, and turning north, made a second landfall over Newfoundland.

Figure 1a shows that simulations of Gustav's storm track are in good agreement with the National Hurricane Center (NHC) analysis track. Gustav moved slowly during the initial phase of the simulation, and accelerated during its extratropical phase as it passes Nova Scotia. Surface currents induced by Gustav are also presented in Fig. 1a, achieving peak speeds to the right of the storm track. Results for SST are shown in Fig. 1b. By 48 h in the MC2–POM-spray simulation, widespread

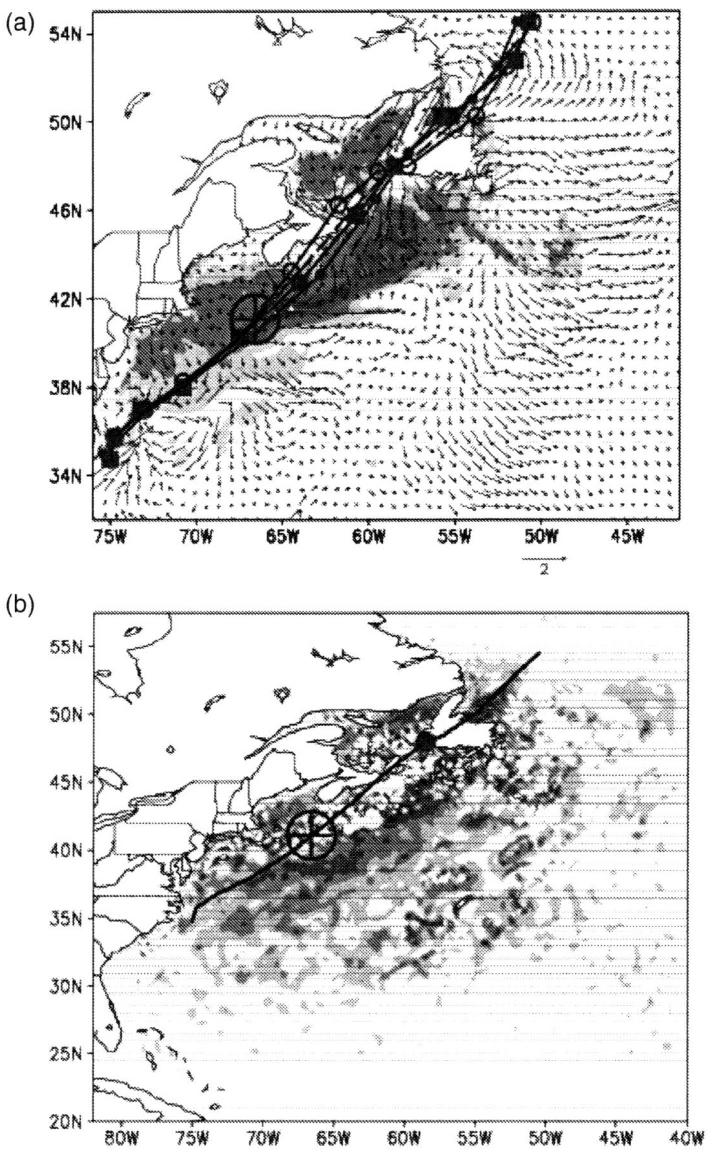


Figure 1: (a) Simulations of Gustav's storm track from 18 UTC September 10 to 12 UTC September 13: control MC2 (—○), MC2–POM (– –), MC2–POM-spray (—■), and NHC analysis (—●). Six-hourly intervals shown. Surface currents at 48 h in the simulations (→ m $s^{-1}$), and SST (°C) at 48 h, minus the initial SST, from MC2–POM-spray model. (b) Difference in SST (°C) distributions of the MC2–POM-spray minus MC2–POM simulations at 48 h. Storm location is ●. Buoy station 44011 is ⊕.

SST cooling occurred on both sides of the storm track, reaching a maximum of approximately −8°C to the right of the storm track. This can be verified by satellite SST data from The Johns Hopkins University (http://fermi.jhuapl.edu/avhrr), and results from the shallow late-summer North Atlantic mixed layer that tends to develop. Regarding spray, maximum differences in surface currents and SST between MC2–POM-spray and MC2–POM simulations are 0.1 m s$^{-1}$ and 0.3°C.

Model verification is given in Fig. 2a–c, comparing observed $U_{10}$, central sea level pressure (SLP), and SST data from National Data Buoy Center (NDBC) buoy 44011 (at 66.59°W, 41.09°N) (see Fig. 1a and b), with coupled and uncoupled model simulations. Although SST cooling appears in the NDBC data and the MC2–POM-spray run, it is not present in the uncoupled MC2 run, which assumes invariant SST values (Fig. 2c). Both the MC2 and the MC2–POM-spray runs compare well to $U_{10}$ and the central SLP buoy data (Fig. 2a and b), suggesting that

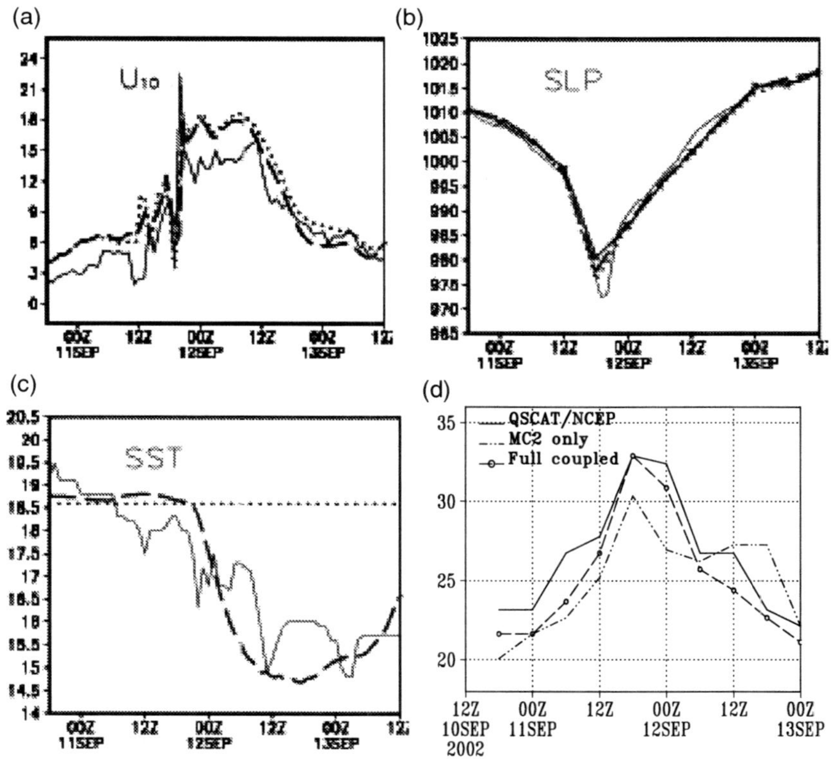


Figure 2: Comparisons of coupled (——) and uncoupled (- - -) model simulations to buoy 44011 data (66.59°W, 41.09°N) for (a) $U_{10}$ (m s$^{-1}$), (b) central SLP, and (c) SST. (d) Comparison of model runs with QSCAT/NCEP data (—), for peak $U_{10}$ following the storm center.

the coupling dynamics has little impact on the storm at this phase of its life cycle. However, at the storm's peak intensity, the spray's impact is as much as about 3.5 m $s^{-1}$ and covers a large area, as suggested in Fig. 1a and b. Perrie *et al.* [21] estimate that the maximum impact of MC2–POM coupling, following Gustav's trajectory, gives 24% less total (latent + sensible) heat than the uncoupled MC2 simulation, and 10% less than the MC2–POM-spray simulation. This reflects greater SST cooling and less storm intensity in the MC2–POM run. Thus, heat fluxes are reduced by SST cooling, but enhanced by spray, particularly in high winds and air–sea temperature differences. Figure 2d compares the coupled model simulation of maximal $U_{10}$ and the observed satellite QSCAT/NCEP blended winds (http://dss.ucar.edu/datasets/ds744.4/), following the storm's trajectory.

Because of high sea state, no $fCO_{2w}$ data were collected when Gustav was closest (about 120 km) to the Ron Brown. Following Bates *et al.* [1], we assume that $fCO_{2w}$ increases linearly from 377 μatm on 06 UT September 11 until about Gustav's peak, and then stays constant at 387 μatm until 14 UT September 12 (Fig. 3). After Gustav's passage, $fCO_{2w}$ and SST oscillated strongly, as the Ron Brown moved through the Gulf Stream. Minor $fCO_{2a}$ oscillations reflect SLP variations. Perrie *et al.* [7] show that model estimates for the storm-enhanced vertical velocity *w*, SST, and temperatures, averaged over $200 \times 200\ km^2$ encompassing the Ron Brown's ship track, suggest oscillatory patterns, of the order of 0.0001 m $s^{-1}$ for *w*, and 2–3°C for SST cooling. These are relatively weak, because the Ron Brown was to the left of the storm track, and rather distant from the area of Gustav's maximum intensity, where *w* and SST cooling reached $\sim 3.5 \times 10^{-4}$ m $s^{-1}$ and ~8°C, respectively, and the thermocline deepened from 15 m initially to a final 30 m.

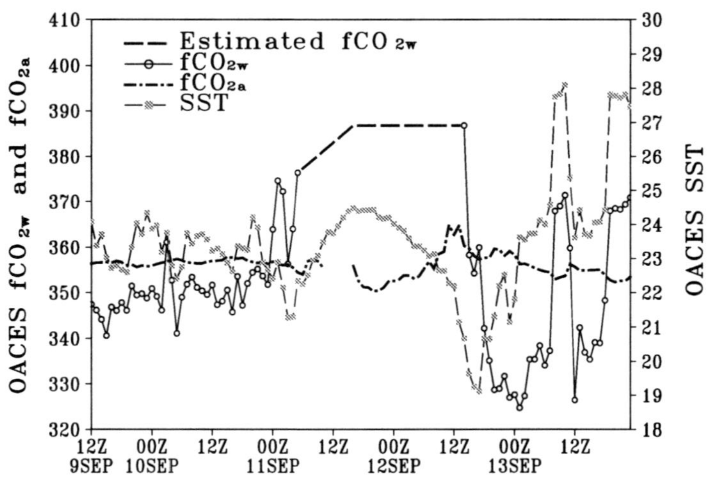


Figure 3: Time series of observed $fCO_{2w}$, $fCO_{2a}$, SST (°C), and estimated $fCO_{2w}$ (- - -).

## 4 Air–sea gas transfer velocity

Using variables from the coupled model simulation, we can estimate the gas transfer velocity $k_L$ and the air–sea $CO_2$ flux, written as

$$Q = k_L \alpha \Delta f CO_2 \tag{4}$$

where $\alpha$ is the solubility of $CO_2$ and $\Delta f CO_2$ is the difference between $f CO_{2w}$ and $f CO_{2a}$. Units for $Q$ are mmol $CO_2$ $m^{-2}$ $h^{-1}$. The simplest $k_L$ relations are given by Wanninkhof [8] and Wanninkhof and McGillis [9], (hereafter "Wanninkhof 92" and "Wanninkhof 99"), respectively,

$$k_L = 0.31 U_{10}^2 (S_c/660)^{-1/2} \tag{5}$$

$$k_L = 0.0283 U_{10}^3 (S_c/660)^{-1/2} \tag{6}$$

in terms of the Schmidt number $S_c$. Units for $k_L$ are cm $h^{-1}$. A third $k_L$ formulation by Zhao *et al.* [10] (hereafter "Toba–Zhao"), uses a wave-breaking parameter $R_B$, and the wave spectrum's peak frequency $\omega_p$,

$$k_L = 0.13 R_B^{0.63} \quad \text{and} \quad R_B = u_*^2/\upsilon\omega_p, \tag{7}$$

where $u_*$ is the air-side friction velocity and $\upsilon$ is the kinematic viscosity.

A fourth $k_L$ formulation (hereafter "Fairall"), by Fairall *et al.* [11], matches water and air fluxes, includes turbulent and radiative fluxes, $U_{10}$, sea state, currents, SST, whitecap fraction $f$, and near-surface thermal structure,

$$k_L = k_{L(\text{bulk})} + k_{\text{breaking}}, \tag{8}$$

where the wave-breaking term $k_{\text{breaking}}$ (Woolf [22]) is

$$k_{\text{breaking}} = f \mathbf{V} \alpha^{-1} [1 + (\boldsymbol{e} \alpha S_c^{-1/2})^{-1/n}]^{-n}, \tag{9}$$

and the whitecapping fraction (Monahan and Torgeresen [23]) is

$$f = 3.8 \times 10^{-6} U_{10}^{\beta}, \quad \beta \approx 3.4, \tag{10}$$

and $\mathbf{V}$, $\boldsymbol{e}$ and $n$ are empirical constants from the GasEx-1998 experiment (see Fairall *et al.* [11]), respectively, 14, 1.2, and 4900 cm $h^{-1}$, and may need readjustment for other field data (Hare *et al.* [24]). The interfacial term $k_{L(\text{bulk})}$ is

$$k_{L(\text{bulk})} = \frac{u_*}{\sqrt{\rho_w/\rho_a}\left(h_w S_{cw}^{1/2} + \ln(z_w/\delta_w)/\kappa\right) + \alpha\left(h_a S_{ca}^{1/2} + C_d^{-1/2} - 5 + \ln(S_{ca})/(2\kappa)\right)}, \tag{11}$$

where the subscript "a" ("w") denotes air (water) side, $\rho$ is the density, $z$ is the measurement depth, $\delta$ is the turbulent surface layer thickness, $\kappa$ is the von Kármán constant, $C_d$ is the drag coefficient, and $h$ is defined by $h \equiv \Lambda R_r^{1/4}/\varphi$, where $\Lambda$ is an adjustable constant, $R_r$ is the roughness Reynolds number, and $\varphi$ is an empirical function that accounts for buoyancy effects on turbulent transfer in the ocean. Variables $u_*$, $C_d$, and $R_r$ come from the Toga-COARE bulk flux parameterizations (Fairall *et al.* [25]). The Schmidt number in air $S_{ca}$ is from table 1 in Fairall *et al.* [11].

The Fairall and Toba–Zhao formulations are very sensitive to changes in $\beta$ and $\omega_p$, respectively, and probably represent the upper bounds on $k_L$. Decreasing $\omega_p$ from $1\,s^{-1}$ to $3\,s^{-1}$, or increasing $\beta$ from 3.4 to 3.6, results in rather large $k_L$ changes. Moreover, whitecap data are extremely noisy, and this variation in $\beta$ does not capture either the data variability or the probable dependence on $U_{10}$ and sea state. The Toba–Zhao $k_L$ was calibrated in freshwater in whitecaps-generated shoaling waves, which can differ from whitecap-related $CO_2$ transfer in seawater (Asher *et al.* [26]).

## 5 Gas transfer velocity and $CO_2$ air–sea flux

Figure 4 compares the $k_L$ formulations at the Ron Brown location. All $k_L$s show the storm's impact. The Wanninkhof 92 relation is systematically the lowest at Gustav's peak, whereas the Toba–Zhao and Wanninkhof 99 $k_L$s are similar, except for wave-related variations. Of the four relations, Fairall's $k_L$ is the highest at Gustav's peak.

Using the four $k_L$ formulations and $fCO_{2w}$ and $fCO_{2a}$ data at the Ron Brown, Fig. 5 estimates the net air–sea $CO_2$ flux $Q$ area-averaged over (71.5°W–73.5°W, 38.5°N–40.5°N) around the Ron Brown. Within hours of Gustav's passing, $Q$ increases to a maximum out-flux of $2.1\,\text{mmol}\,m^{-2}\,h^{-1}$ (assuming Fairall's $k_L$), decreases to 0 within 18 h, and briefly to $-0.4\,\text{mmol}\,m^{-2}\,h^{-1}$, thereafter.

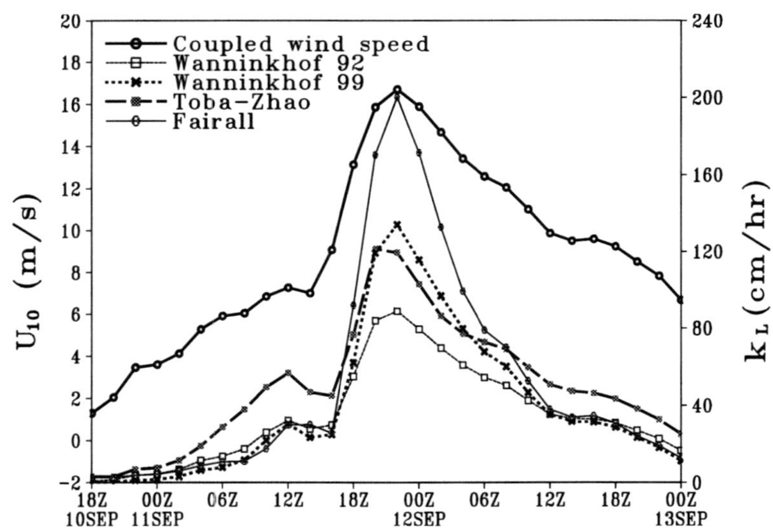


Figure 4: Time series of $k_L$ formulations, averaged over area (71.5°W–73.5°W, 38.5°N–40.5°N), encompassing the Ron Brown.

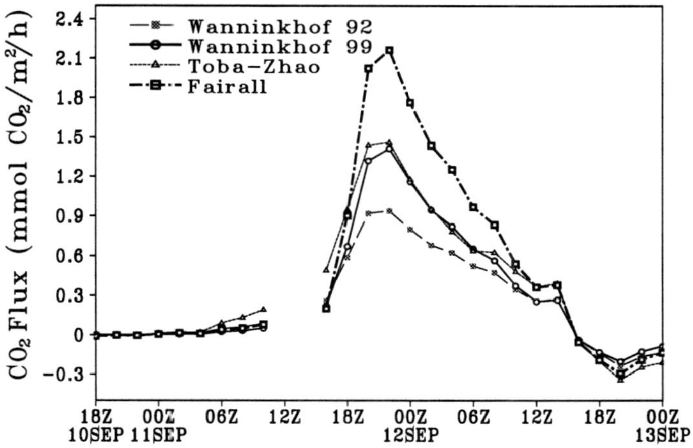


Figure 5: Time series of $CO_2$ flux $Q$, using four $k_L$ formulations, averaged over (71.5°W–73.5°W, 38.5°N–40.5°N) around the Ron Brown. The gap at 12 UT occurs because no $fCO_{2a}$ data is available.

## 6 Concluding remarks

OACES data was used to estimate the impact of Hurricane Gustav on air–sea $CO_2$ exchange, comparing four recent $k_L$ formulations. We simulated Hurricane Gustav using a coupled atmosphere–ocean wave-spray model. Model estimates were shown to agree well with *in situ* and analysis data from CMC and NHC, including storm-induced SST cooling, resulting from strong upwelling and entrainment mixing below the mixed layer. In high winds, the Wanninkhof 92 $k_L$ is the lowest in magnitude, whereas the Wanninkhof 99 and Toba–Zhao $k_L$ formulations are similar, although the latter exhibits wave-related variability. During Gustav's peak intensity, Fairall's $k_L$ is the highest in magnitude, more than twice the Wanninkhof 92 $k_L$. Formulations for $k_L$ by Fairall *et al.* [11] and Zhao *et al.* [10] are sensitive to whitecap coverage and peak frequency, respectively. Gustav is estimated to have caused an increased $CO_2$ out-flux, reaching a peak of 2.1 mmol $m^{-2} h^{-1}$, which occurred at the storm's peak and diminished rapidly within 18 h. This result assumes Fairall's $k_L$ and interpolates the $fCO_{2w}$ data across the peak of the storm.

## Acknowledgments

We acknowledge funding from the Canada Panel on Energy Research and Development, Natural Science and Engineering Research Council and Canadian Foundation for Climate and Atmospheric Sciences (SOLAS), and assistance with the MC2 model (J. Gyakum, R. McTaggart-Cowan), the sea-spray model (E. Andreas), the URI POM version (I. Ginis), and the fourth $k_L$ model (C. Fairall).

## References


[1] Bates, N.R., Knap, A.H. & Michaels, A.F., Contribution of hurricanes to local and global estimates of air-sea exchange of $CO_2$. *Nature*, **395**, pp. 58–61, 1998.

[2] Ren, X., Perrie, W., Long, Z., Gyakum, J. & McTaggart-Cowan, R., On the atmosphere-ocean coupled dynamics of cyclones in midlatitudes. *Monthly Weather Review*, **132**, pp. 2432–2451, 2004.

[3] Chan, J.C.L., Duan, Y. & Shay, L.K., Tropical cyclone intensity change from a simple ocean-atmosphere couple model. *Journal of the Atmospheric Sciences*, **58**, pp. 154–172, 2001.

[4] Jacob, S.D., Shay, L.K., Mariano, A.J. & Black, P.G., The 3D oceanic mixed layer response to hurricane Gilbert. *Journal of Physical Oceanography*, **30**, pp. 1407–1429, 2000.

[5] Bender, M.A. & Ginis, I., Real-case simulations of hurricane-ocean interaction using a high-resolution coupled model: effects on hurricane intensity. *Monthly Weather Review*, **128**, pp. 917–946, 2000.

[6] Schade L.R. & Emanuel, K.A., The ocean's effect on the intensity of tropical cyclones: results from a simple coupled atmosphere-ocean model. *Journal of the Atmospheric Sciences*, **56**, pp. 642–651, 1999.

[7] Perrie, W., Zhang, W., Ren, X. & Long, Z., The impact of Hurricane Gustav on air-sea $CO_2$ exchange. *Geophysical Research Letters*, **31**, L09300, doi: 10.1029/2003GL019212, 2004.

[8] Wanninkhof, R., Relationship between wind speed and gas transfer exchange over ocean. *Journal of Geophysical Research*, **97**, pp. 7373–7382, 1992.

[9] Wanninkhof, R. & McGillis, W.R., A cubic relationship between gas transfer and wind speed. *Geophysical Research Letters*, **26**, pp. 1889–1892, 1999.

[10] Zhao, D., Toba, Y., Suzuki, Y. & Komori, S., Effect of wind waves on air-sea gas transfer: proposal of an overall $CO_2$ transfer velocity formula as a function of breaking-wave parameter. *Tellus*, **55B**, pp. 478–487, 2003.

[11] Fairall, C.W., Hare, J.E., Edson, J.B. & McGillis, W.R., Parameterization and micro-meteorological measurements of air-sea gas transfer. *Boundary-Layer Meteorology*, **96**, pp. 63–105, 2000.

[12] Zhang, W., Perrie, W. & Li, W., The role of waves and sea spray in mid-latitude storms. Accepted in *Monthly Weather Review*, 2005.

[13] Benoit, R., Desgagne, M., Pellerin, P., Chartier, Y. & Desjardins, S., The Canadian MC2: a semi-implicit semi-Lagrangian wide-band atmospheric model suited for fine-scale process studies and simulation. *Monthly Weather Review*, **125**, pp. 2382–2415, 1997.

[14] Andreas, E.L., An algorithm to predict the turbulent air-sea fluxes in high-wind, spray conditions. *12th Conf. Interaction of the Sea and Atmosphere*, Long Beach, CA, American Meteorological Society, 2003.

[15] Andreas, E.L. & DeCosmo, J., The signature of sea spray in the HEXOS turbulent heat flux data. *Boundary-Layer Meteorology*, **103**, pp. 303–333, 2002.

[16] Fairall, C.W., Bradley, E.F., Rogers, D.P., Edson, J.B. & Young, G.S., Bulk parameterization of air-sea fluxes for Tropical Ocean-Global Atmosphere Coupled-Ocean Atmosphere Response Experiment. *Journal of Geophysical Research*, **101**, pp. 3747–3764, 1996.

[17] Mellor, G.L., User's guide for a three-dimensional primitive equation numerical ocean model. Program in Atmospheric and Oceanic Sciences, Princeton University. http://www.aos.princeton.edu/WWWPUBLIC/htdocs.pom, 1998.

[18] Tolman, H.L. & Chalikov, D.V., Source terms in a third-generation wind wave model. *Journal of Physical Oceanography*, **26**, pp. 2497–2518, 1996.

[19] Tolman, H.L., Alleviating the garden sprinkler effect in wind wave models. *Ocean Modelling*, **4**, pp. 269–289, 2002.

[20] Andreas, E.L. & Emanuel, K.A., Effects of sea spray on tropical cyclone intensity. *Journal of the Atmospheric Sciences*, **58**, pp. 3741–3751, 2001.

[21] Perrie, W., Ren, X., Zhang, W. & Long, Z., Simulation of extatropical Hurricane Gustav using a coupled atmosphere-ocean-sea spray model. *Geophysical Research Letters*, **31**, L03110, doi:1029/2003GL018571, 2004.

[22] Woolf, D.K., Bubbles and their role in gas exchange. *The Sea Surface and Global Change*, eds R.A. Duce & P.S. Liss, Cambridge University Press: New York, pp. 173–205, 1997.

[23] Monahan, E.C. & Torgeresen, T., The enhancement of air-sea gas exchange by oceanic whitecapping. *Air-Water Mass Transfer: Proc. 2nd Int. Symp. Gas Transfer at Water Surfaces*, eds S.C. Wilhelms & J.S. Gulliver, ASCE: New York, pp. 608–617, 1991.

[24] Hare, J.E., Fairall, C.W., McGillis, W.R., Ward, B. & Wanninkhof, R., Evaluation of the NOAA/COARE air-sea gas transfer parameterization using GasEx data. *Journal of Geophysical Research*, in press, 2005.

[25] Fairall, C.W., Bradley, E.F., Hare, J.E., Grachev, A.A. & Edson, J.B., Bulk parameterization of air-sea fluxes: updates and verification for the COARE algorithm. *Journal of Climate*, **16**, pp. 571–591, 2003.

[26] Asher, W.E., Karle, L.M. & Higgins, B.J., On the differences between bubble-mediated air-water transfer in freshwater and seawater. *Journal of Marine Research*, **55**, pp. 813–845, 1997.

# CHAPTER 7

# Extratropical transition of tropical cyclones in the North Atlantic

J.L. Evans[1] & R.E. Hart[2]
[1]*Department of Meteorology, The Pennsylvania State University, Pennsylvania, USA.*
[2]*Florida State University, Florida, USA.*

## Abstract

Apart from a few early case studies that documented tropical cyclones (TCs) that tracked into the extratropics, it was generally accepted through the 1970s that TCs would decay as they moved out of the tropics. This preconception was challenged in the 1980s. Over the past two decades, a flurry of research on these *extratropically transitioning* TCs has revealed much about their behavior. Here, we discuss extratropically transitioning TCs, beginning from their tropical formation, the conditions under which they evolve, and their spatial and temporal distributions. Case studies are presented and potential indicators of transition, of use to the forecast community, are introduced.

Case studies demonstrate the need for research into the interactions between the evolving TC and an approaching higher latitude trough. Constructive interactions between these systems, when accompanied by a supportive remnant tropical environment, will lead to transition and possibly reintensification of the storm as an extratropical system.

Key results include a late season maximum in the percentage of storms that undergo transition. This late season peak is related to the need for synoptic support for tropical development to intensify cyclones prior to transition, followed almost immediately by a strong baroclinic synoptic energy source as transition occurs. Recognition of key structural changes as the storm transitions has led to the development of a number of indicators of transition currently being tested by the forecast community.

## 1 Introduction

Extratropical transition (or ET) is the process by which a tropical cyclone (TC) is transformed into an extratropical cyclone (EC). These EC systems often persist as coherent structures far into the extratropics. In the Atlantic, systems with tropical beginnings have been tracked poleward of 50°N and to the coast of Western Europe, from Spain to Scandinavia (e.g. Bob in 1991, Fig. 1). These systems carry their past with them, as measured by their equivalent potential temperature and potential vorticity signatures. The importance of these latent and dynamic energy sources is explored in the case study example presented in Section 4.

Prior research into ETs in the Atlantic consisted of two papers on a single case in the first half of the 20th century (Tannehill [1]; Pierce [2]), occasional case studies since the 1950s (e.g. Palmén [3]; Kornegay and Vincent [4]; DiMego and Bosart [5, 6]), and a recent rush of studies (e.g. Bosart and Bartlo [7]; Bosart and Lackmann [8]; Thorncroft and Jones [9]; Hart and Evans [10]; Evans and Hart [11]; Jones *et al.* [12]; Atallah and Bosart [13]; Arnott *et al.* [14]). The history of ET research in the Pacific basin is similar (e.g. Sekioka [15, 16, 17]; Matano and Sekioka [18]; Brand and Guard [19]; Harr and Elsberry [20]; Harr *et al.* [21]; Klein *et al.* [22]; Sinclair [23]), with even less attention being paid to the Indian Ocean (Foley and Hanstrum [24]).

Poleward movement of a TC is expected to lead to the decay of the system as it encounters the hostile, strongly sheared environment and cooler waters (or land) of the midlatitudes. The exceptions to this rule are extratropically transitioning TCs. Cyclones undergoing ET may initially weaken as they move into the extratropics, but these systems can reintensify into midlatitude storms that are

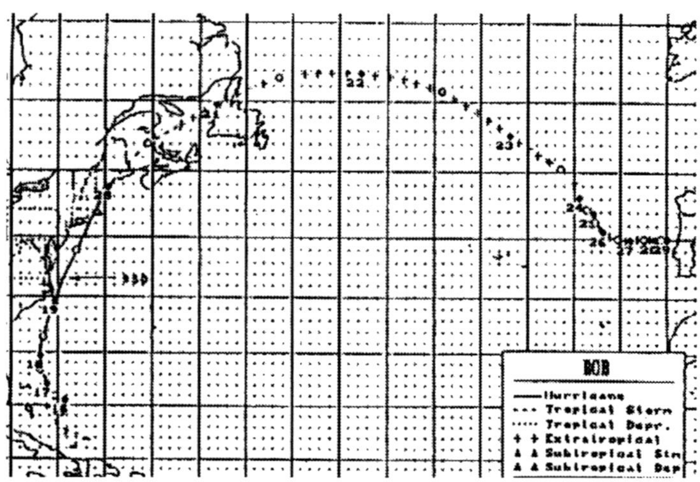


Figure 1: Best track of Hurricane Bob (1991) from August 16–29, 1991. This storm greatly impacted the United States, Canada, and Western Europe. 0000 UTC (1200 UTC) positions are marked with solid (open) circles.

occasionally more intense than their tropical selves (e.g. Irene 1999, Prater-Mayes and Evans [25]; Evans and Prater-Mayes [26]; Agusti-Panareda *et al.* [27]), often traveling at forward speeds of 15–20 m s$^{-1}$. More commonly, the remnant tropical storm can provide a source of enhanced thermal contrast for the later development of an intense midlatitude storm (e.g. Abraham *et al.* [28]). These systems are often poorly forecast by numerical forecast models (e.g. Rabier *et al.* [29]) and can visit severe weather on the mid-Atlantic and New England coasts of the United States and the Canadian Maritimes, as well as Western Europe (including Scandinavia). Their rapid forward speed, the large size of their gale force wind area and intense asymmetric rainfall region, and the ocean waves they generate provide difficult forecast challenges to the responsible forecast centers around the globe. These intense weather phenomena can exist long after their "tropical storm" status has been discontinued (Abraham *et al.* [28]).

The rapid movement of a transitioning storm and its large area of high wind speeds combine to generate large ocean surface waves and swell (Bowyer [30]; Moon *et al.* [31]). Indeed, MacAfee and Bowyer [32, 33] suggest that the rapid forward speeds of ET storms can result in a "trapped fetch" phenomena in which the storm moves with the fastest growing waves it is generating, providing a much longer period of sustained forcing and resulting in much larger waves than would be possible if the storm were moving at only a few meters per second (as for TCs). Hence, the waves generated by an extratropically transitioning TC can be much larger than the waves generated by a more intense, but slower moving, TC. Enhancement in wave growth occurs only to the right of the track (northern hemisphere; left in the southern hemisphere), in the fastest wind region of the ET storm (where the storm's forward motion is in the same direction as its winds). Bowyer [30] demonstrated this trapped fetch phenomena for the case of Hurricane Luis (1995). Peak waves of over 30 m were generated by Luis and caused extensive damage to the QEII cruise ship. Moon *et al.*'s [31] simulation of detailed wave-wind coupling mechanisms demonstrated skill in modeling the two-dimensional spectra associated with 14 + m significant wave heights produced by the transitioning Hurricane Bonnie (1998) and recognized the challenges in simulating shoaling waves as a storm approaches landfall.

In this chapter, we review the early research and recent advances in the understanding of ET in the North Atlantic basin. We begin in Section 2 with a review of formation (or "genesis") of TCs and their basic structure. A climatology of TCs and ET in the North Atlantic is presented in Section 3. This is followed by an examination of intraseasonal and interbasin variations in ET in Section 4, in which we propose favorable environments for transition. The percentage of tropical storms that undergo ET, that reintensify in their extratropical phase, and those that make landfall are identified. We explore the origins of these systems and their paths to transition. A "typical" ET case study, Hurricane Floyd (1999), is documented in Section 5. In Section 6 we discuss the informal set of criteria that have evolved for diagnosing and forecasting ETs and we introduce an alternative objective approach, the cyclone phase space (CPS) (Hart [34]; Evans and Hart [11]). A brief summary is provided in Section 7.

## 2 A basic primer on TC formation and structure

According to the World Meteorological Organization (WMO), a TC is a warm-core, nonfrontal, low-pressure system that develops over the ocean and has a definite organized surface circulation. A tropical storm is a TC whose intensity (measured as the maximum sustained surface wind speed) exceeds $63\,\text{km}\,\text{h}^{-1}$ (39 mph). In the Atlantic basin, intensity is measured by the peak *one-minute* averaged wind near the center, although the convention followed in the western Pacific and Indian Oceans is for *ten-minute* averaged winds to be used for intensity. A TC is assigned a name once it reaches tropical storm status (according to the local wind averaging convention). TCs whose intensity exceeds $119\ \text{km}\ \text{h}^{-1}$ (74 mph) are classified as hurricanes in the Atlantic and eastern Pacific basins and as typhoons in the western North Pacific. Elsewhere, they are simply known as "severe tropical cyclones".

TCs are a natural part of the tropical climate system, occurring in all tropical ocean basins, with the exception of the South Atlantic (Fig. 2), although rare storms in the South Atlantic with tropical characteristics are observed. TCs have impacted the course of history, scattering and damaging Christopher Columbus' fleet in his fourth Caribbean voyage (and sinking the Spanish treasure fleet) in 1502, and thwarting the attempts of Kublai Khan to invade Japan in 1266 (Hatada [35]; Yanai [36]). As a result, TCs have inspired awe, curiosity, and, more recently, research.

Due to the differing sense of cyclonic flow across the equator, TC wind and cloud patterns are reflected for storms north and south of the equator: TCs spin counterclockwise in the northern hemisphere and clockwise in the southern hemisphere, with corresponding variations in their spiral rainband structure. In a TC, the wind flows inward cyclonically at lower levels, spiraling upward in the zones of deep convection (the central eyewall or the spiral rainbands), ultimately spiraling outward aloft, just below the tropopause (Fig. 3). Emanuel [38] closes this cycle in his theoretical model of the TC as a Carnot engine by hypothesizing a sinking branch of air far from the storm center, which ultimately recycles as inflowing air

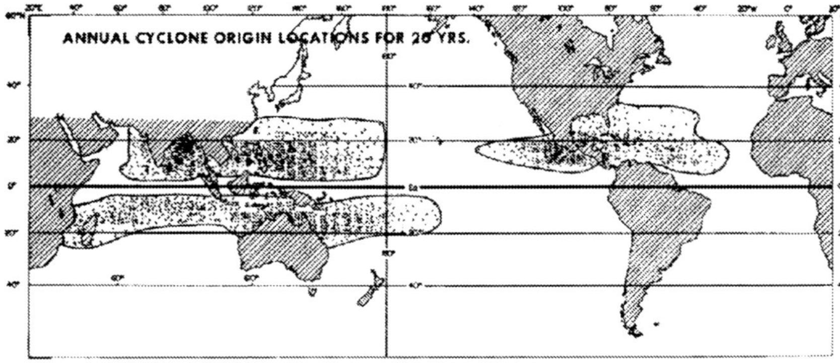


Figure 2: Global distribution of tropical cyclogenesis locations for a 20-year period (reproduced from Gray [37]).

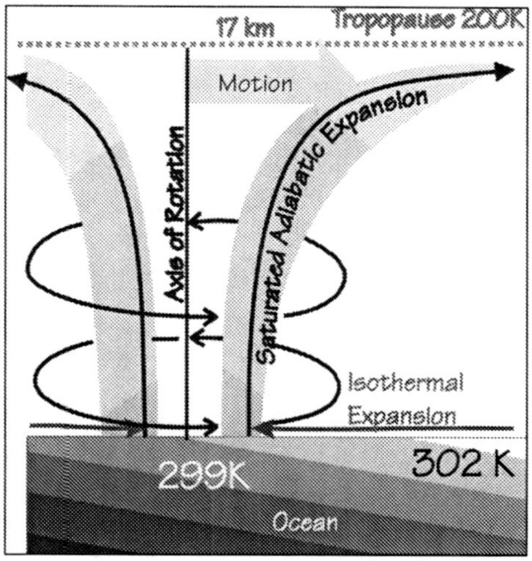


Figure 3: Schematic of the three-dimensional wind flow patterns in a typical TC (Hugh Willoughby, 2002, personal communication).

near the surface. Inflowing air is approximately isothermal, but increases its moist entropy via evaporation at the surface and decreasing pressure as it approaches the storm center.

The pattern of moist convection evident from satellite imagery is one of the most recognizable features of a TC. Four infrared satellite images for Hurricane Floyd (1999) depict the storm structure at different stages of its life cycle (Fig. 4). The clear region in the center of a mature tropical storm is known as the eye and is relatively calm with light winds and the lowest surface pressure. An organized band of thunderstorms immediately surrounds this calm center: this is the eyewall and the strongest winds are to be found on the inner flank of this thunderstorm annulus (Fig. 5).

## 3 Atlantic TC and ET climatology

An average of 10 named tropical storms (five hurricanes) form in the North Atlantic annually. As few as four tropical storms [in 1983] (two hurricanes [in 1982]), and as many as 21 tropical storms [in 1933] (12 hurricanes [in 1969]) have been recorded in individual years (Neumann *et al.* [40]). The earliest tropical storm formation occurred in April 2003, when Tropical Storm Ana evolved from an initially subtropical cyclone, although the hurricane season in the Atlantic, Caribbean, and Gulf of Mexico runs from June 1 to November 30. In addition, the 2003 season provided multiple landfalls (Fabian, Isabel, Juan) and ended with a rare December storm (TC Peter).

Gray [41, 42] documented six necessary (but not sufficient) conditions for tropical cyclogenesis: (i) ocean temperatures exceeding 26.5°C to a depth of 60 m; (ii) moist mid-troposphere; (iii) conditional instability; (iv) enhanced low-level

(a)

(b)

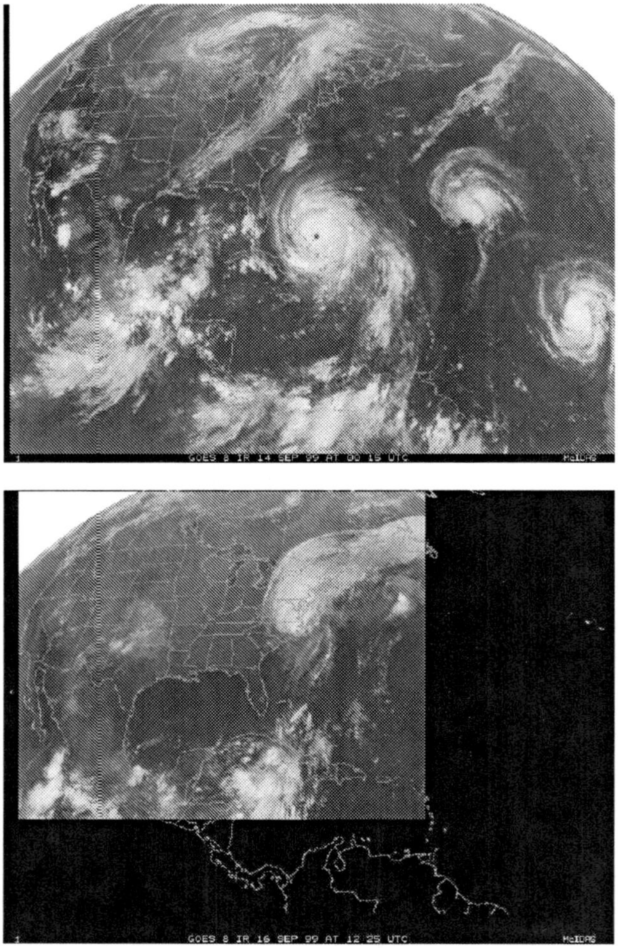


Figure 4: Hurricane Floyd (1999) GOES-8 infrared satellite imagery: (a) 0000 UTC September 14 (near peak tropical intensity), (b) 1200 UTC September 16 (ET commenced), (c) 1200 UTC September 17 (ET completed), and (d) 0000 UTC September 19 (EC over Canada).

relative vorticity; (v) weak vertical shear of the horizontal winds between 850 and 300 hPa in the genesis region, with enhanced easterly shear poleward and westerly shear equatorward; (vi) location at least 5° latitude from the equator. The seasonal evolution of these parameters explains the August through October peak in genesis frequency: the vertical wind shear tends to be weaker earlier in the season, whereas the thermal inertia of the ocean means that higher ocean temperatures and a more conditionally unstable atmosphere lag the solar peak and occur later in the season. The requirement for enhanced low-level relative vorticity is equivalent to the need for an incipient disturbance which will reduce the

(c)

(d)

Figure 4 (*continued*)

local Rossby radius of deformation and concentrate the convective heating locally, dropping the surface pressure of the developing system and thereby enhancing the near-surface winds. In the frictional boundary layer, the increase in surface winds results in a concomitant increase in the frictional moisture convergence (whether by locally enhanced surface evaporation or radial import) which feeds the deep convection in a positive feedback cycle. Incipient disturbances may be African easterly waves (Molinari *et al.* [43]), other equatorial and near-equatorial waves (Dickinson and Molinari [44]), mesoscale convective complexes (Bister and Emanuel [45]) or subtropical low pressure systems (Bosart and Bartlo [7]). The Madden Julian Oscillation modulates the organized convection in the Atlantic basin (Evans and Jaskiewicz [46]), contributing to the intraseasonal variability of tropical cyclogenesis.

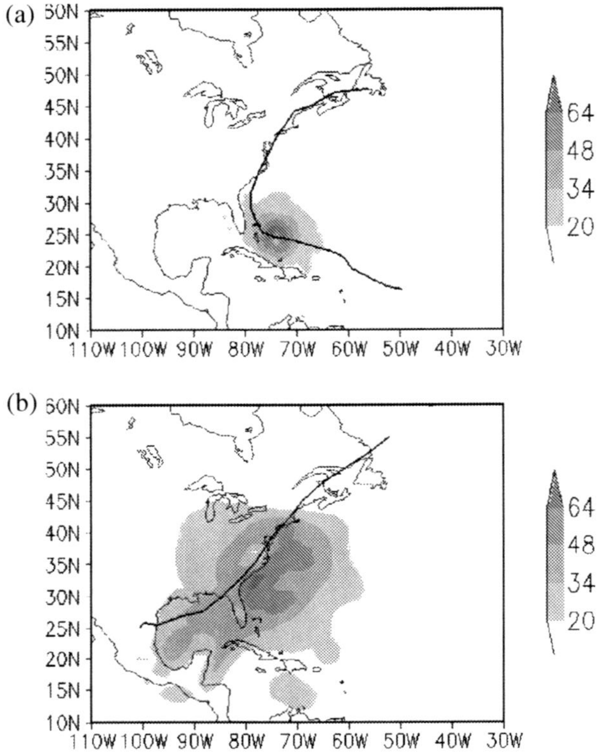


Figure 5: Comparison of 10 m wind speed distributions in examples of a typical TC and EC. Pictured are (a) Hurricane Floyd at 0000 UTC September 14, 1999 [NOGAPS 1° analysis]; (b) the "Blizzard of 1993" at 1200 UTC March 13, (NCEP/NCAR 2.5° reanalysis; Kalnay *et al.* [39]). Wind speed shading (light to dark) corresponds to winds in excess of 20 kt (37 km $h^{-1}$), 34 kt (63 km $h^{-1}$), 48 kt (89 km $h^{-1}$), and 67 kt (124 km $h^{-1}$). Note the expanded maximum wind radius of the EC (b) compared with that of the TC.

Forty-six percent of the named Atlantic tropical storms ultimately undergo ET, as defined by the National Hurricane Center (NHC) (Hart and Evans [10]). Since 1950, up to 70% and as few as 30% of these storms in a given season have been observed to undergo ET, with a standard deviation of 12%. Linear correlations with several conventional measures of climatological variability in the North Atlantic (PNA, NAO, SOI) individually describe no more than about 10% of the observed year-to-year variability. Thus, the likelihood of ET in a particular season is not readily tied to one of these measures of teleconnections. This suggests that local effects, or perhaps spatial and temporal phasing between independent systems, play a major role in determining the individual storm's potential for transition (Hart and Evans [10]).

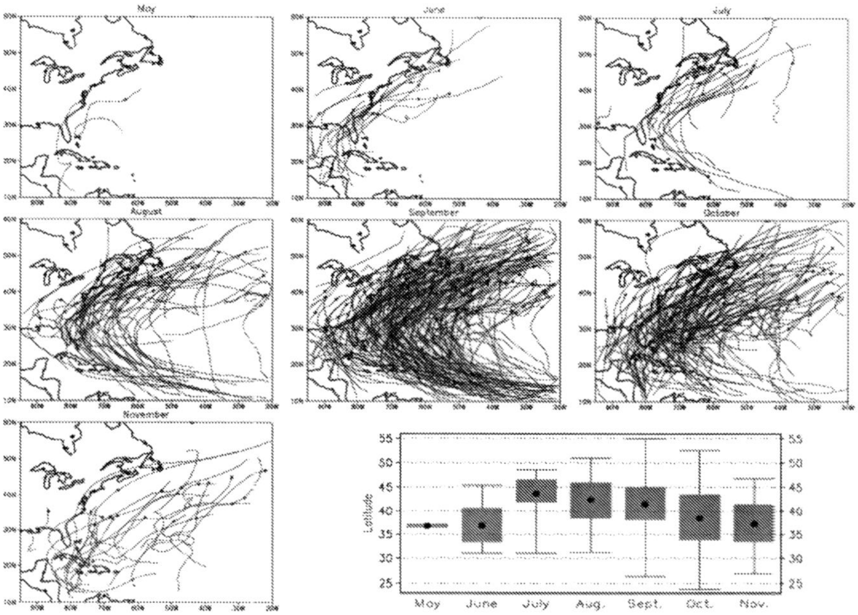


Figure 6: Tracks of extratropically transitioning North Atlantic TCs by month. Dots represent the point of transition as defined by NHC. The monthly range of latitudes over which ET has been observed is depicted in the bar chart: solid circles denote the mean transition location for each month; lighter bars represent the first and third quartiles of the transition latitudes; the horizontal bars beyond these are the extreme latitudes of observed transition. Reproduced from Hart and Evans [10].

Extratropically transitioning TCs represent almost 50% of landfalling TCs on the east coasts of the United States and Canada, and the west coast of Europe, combined (Hart and Evans [10]). The likelihood of ET increases through the tropical storm season, with a peak of 50% in October (48% in September and less than 40% in all other months) (Fig. 6). The coastal Atlantic areas most likely to be impacted by a transitioning TC are the northeast United States and the Canadian Maritimes (1–2 storms per year), and Western Europe (once every 1–2 years) (Fig. 7).

Hart and Evans [10] evaluated storm intensity changes during and after ET of the 61 transitioning tropical storms in 1979–1993. While it is extremely rare for a transitioning TC to regain its peak tropical intensity in its extratropical phase (Fig. 8a and b), 51% of ET cases underwent post-transition intensification. Over 60% of cyclones that underwent post-transition intensification originated south of 20°N (Fig. 8c), although only 52% of Atlantic storms from 1948–2000 formed in this zonal belt. In contrast, 90% of TCs that underwent post-transition decay originated north of 20°N (Fig. 8d), suggesting that strong baroclinic characteristics during formation hinder post-transition intensification and shorten the post-transition

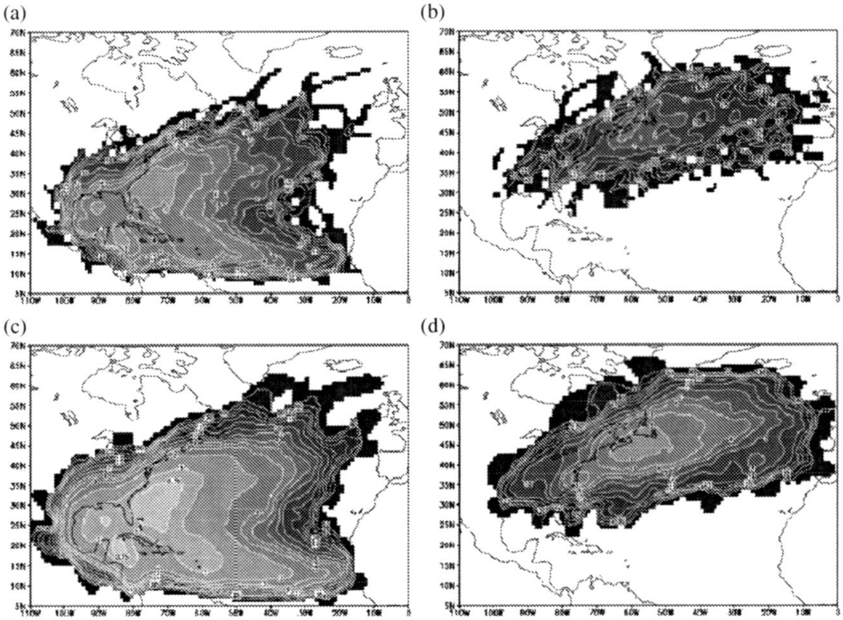


Figure 7: Mean return period in years for passage of a storm within a chosen distance of a location: (a) TC within 110 km (60 n/mi); (b) ET within 110 km; (c) TC within 300 km. (d) ET within 300 km. Figures are based on NHC best tracks from 1899–1996. Reproduced from Hart and Evans [10].

life of the ET system. Some environmental factors that modulate both the peak intensity and the likelihood of transition are explored in the next section.

## 4 Intraseasonal and interbasin variation in ET and proposed favorable environments

Expressed as a percentage of annual tropical storms, the Atlantic Ocean is the basin in which ET is most frequent: 46% of Atlantic TCs undergo ET (Hart and Evans [10]), while only 27% of western North Pacific storms transition. However, the much higher incidence of tropical storms in the western North Pacific basin makes it the most frequent location for ET, as noted by Klein *et al.* [22]. In the South Indian Ocean (Foley and Hanstrum [24]) and South Pacific (Sinclair [23]), ETs occur relatively infrequently, involving only 10% of tropical storms, and is not recorded in the North Indian Ocean (e.g. Grandau and Engel [47]).

Hart and Evans [10] documented the seasonal cycle of transition location and frequency in the Atlantic. They attribute this intraseasonal variation in ET latitude as the result of competing support for TC and EC cyclone development. ET occurs over the latitude range 24°N–55°N, with the highest incidence between

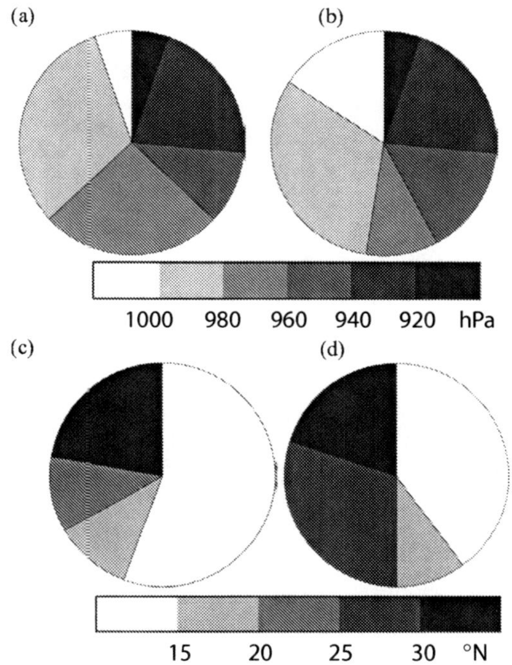


Figure 8: Characteristics of storms that go on to transition: pie charts of (a) peak tropical intensity and (b) peak post-transition intensity. Genesis latitudes for (c) reintensifying storms and (d) weakening storms.

35°N–45°N (Fig. 6). Transition occurs at lower latitudes at the beginning and end of the season, and at higher latitudes during the tropical season peak in August and September.

The peak seasonal frequency of ET occurs in September and October, when (on average) storm tracks are predominantly confined to environments that are either tropically or baroclinically supportive for development (Fig. 9). The delayed warming of the Atlantic Ocean allows the location of transition to shift northward late in the season, since the critical threshold for tropical development pushes northward. Conversely, the climatologically favored region for baroclinic development expands southward late in the season, pinching off the oceanic surface area over which tropical development can occur (Hart and Evans [10]). The relative positions of these two areas define the typical life cycle of a transitioning TC: tropical intensification, tropical decay, extratropical transition and intensification, occlusion. This seasonal tension between development regions was demonstrated using ECMWF reanalysis data for the 1985 season by Hart and Evans [10]. Global maps of these analyses have been calculated using the NCEP 2.5° × 2.5° reanalyses data of Kalnay *et al.* [39] (Fig. 9).

The intraseasonal variation evident in the North Atlantic is reproduced in the western North Pacific and is hinted at in the North and South Indian Oceans.

(a)

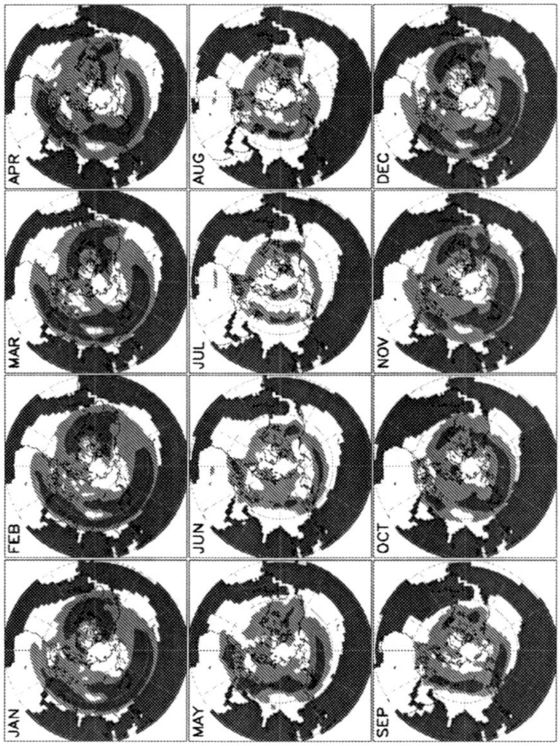


Figure 9: Comparison of the monthly mean distribution of geographic regions supporting tropical and baroclinic storm development in (a) the northern hemisphere and (b) the southern hemisphere. Regions of SST > 28°C have dark shading; light (medium) shading indicates regions in which the Eady baroclinic growth rate of the most unstable mode is greater than $0.25\,\text{day}^{-1}$ ($0.5\,\text{day}^{-1}$). Fields are calculated from 1982–2001 NCEP/NCAR 2.5° × 2.5° monthly mean reanalyses generated by Kalnay *et al.* [39], and so probably underestimate extreme values.

The potential for ET in the North Indian Ocean (Bay of Bengal, Fig. 9a) is strongest in March, April, and November, consistent with the dual TC seasons experienced in this part of the globe. Based on this diagnostic in the southern hemisphere (Fig. 9b), the most likely months for ET off the Western Australian coast are November, April, and May: the transition months for the monsoon season there.

A storm-centered composite of the synoptic environment at the time of ET onset was calculated for a set of 11 cases of North Atlantic ET from 1998–2002. Maps of 300 hPa PV and $\vartheta_e$ (Fig. 10a) demonstrate that the transitioning storm is still distinct from the trough to its west at the time of ET onset. Using the 1PVU threshold as a measure of each system's boundary, it is clear that these systems are less than 10° apart at the time of transition. The TC is still warm cored, but is beginning to distort

(b)

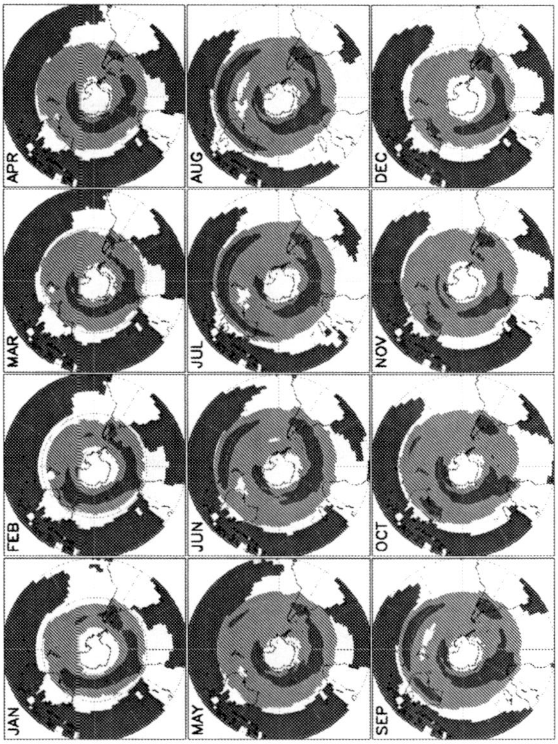


Figure 9 (*continued*)

parallel to the trough axis (NE–SW tilt evident in the storm-centered $\vartheta_e$ contours and shading of 0.5PVU region in Fig. 10a). Comparison of the upper PV and the 1000–500 hPa thickness contours (Fig. 10b) reinforces the independence of the TC and the approaching trough: the TC is still embedded in the tropical environment, although its low-level wind field is clearly beginning to distort. A much broader region of strong winds at 925 hPa is evident to the east of the TC than to its west.

## 5 A typical Atlantic ET case study: Hurricane Floyd (1999)

Hurricane Floyd (1999) was the deadliest US hurricane since Agnes in 1972, with 56 deaths in the United States and one on the Grand Bahama Island attributed to this storm. Floyd was almost a Category 5 hurricane when it passed just to the north and east of the central and northern Bahama islands. Over 50% of the fatalities occurred *after* Floyd made landfall in North Carolina, predominantly due to drowning in freshwater flooding (Lawrence *et al.* [48]). Floyd continued up the coast into the mid-Atlantic and New England regions, now as an extratropically transitioning system. Floyd rained heavily in this region, leaving a further trail of flooding and fatalities in its wake. Total damage estimates for the United States range from

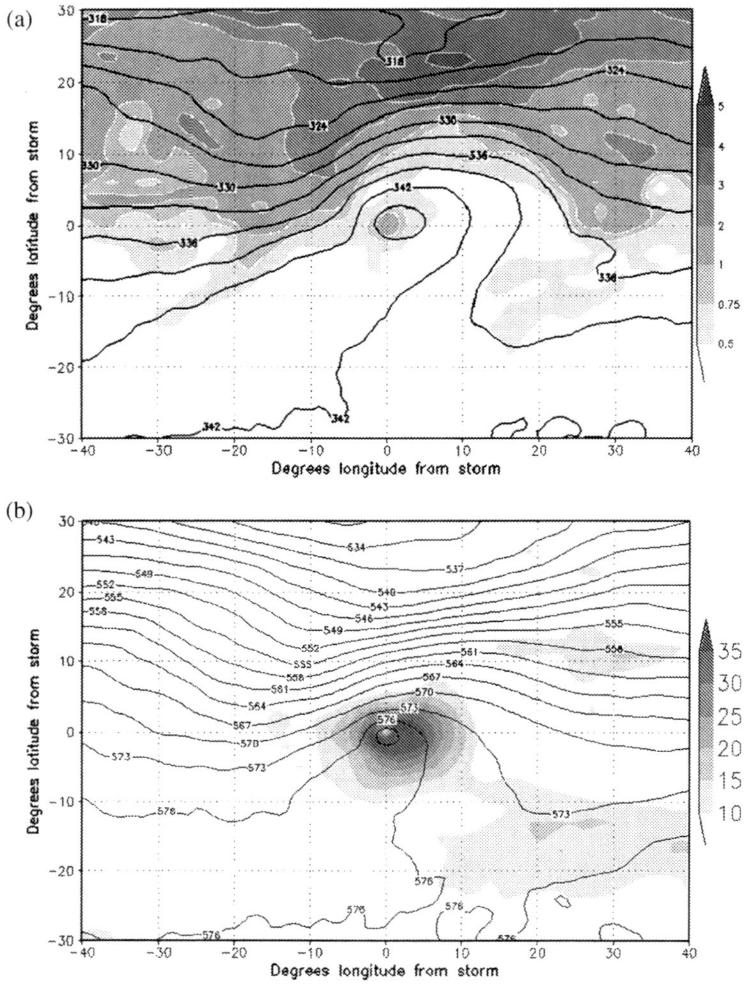


Figure 10: Storm-centered composite synoptic environment at the time of ET onset for 11 cases of North Atlantic ET: (a) 300 hPa PV (PVU; shading PV > 0.25, 0.5, 0.75, 1.0, 2.0, 3.0) and $\vartheta_e$ (contoured every 3 K); (b) 1000–500 hPa thickness (contoured every 3 dm) and 925 hPa wind speed (kt; shaded for: >10 kt, 15 kt, 20 kt, 25 kt, 30 kt, 35 kt). Fields are from the NCEP AVN model analyses.

US$3–6 billion or more (Pasch *et al.* [49]). After being declared extratropical, Floyd continued into the Canadian Maritimes, reintensifying and causing further damage.

The evolution of Hurricane Floyd (1999) from an intense TC into an EC is described here as a typical example of ET. Floyd (1999) represents a case where ET occurs relatively quickly, with a hybrid cyclone phase (frontal warm core) of approximately 24 h, compared with the mean transition period of 33.4 h identified by Evans and Hart [11]. Floyd had its beginning as Tropical Depression Eight, which developed from an unremarkable easterly wave originating in the Cape Verde region. It was identified at 1800UTC on September 7, 1999 and was named Tropical Storm Floyd at 0600UTC on September 8, 1999 (Lawrence *et al.* [48]). At this time, Floyd was located about 75 nmi (1400 km) to the east of the Leeward Islands. As it drifted west–northwest, Floyd slowly intensified, reaching hurricane strength at 1200UTC on September 10, now about 200 nmi (370 km) to the east–northeast of the Leeward Islands (Pasch *et al.* [49]).

A strengthening ridge to the north of Floyd resulted in a turn to the west early on the 12th. In the subsequent 24 h, the central pressure dropped rapidly by about 40 hPa and the maximum sustained winds increased from 95 knots to 135 knots (i.e. 136–250 km $h^{-1}$). Hurricane Floyd reached its peak tropical intensity of 921 hPa (135 kt, 250 km $h^{-1}$) at 1200UTC on September 13, 1999, as a symmetric warm-cored TC and remained an intense Category 4 hurricane for about 12 h (Pasch *et al.* [49]). It made landfall twice in the central and northern Bahamas on September 13 and 14, undergoing an eyewall replacement cycle. Floyd was a large, symmetric tropical system at this time, located well ahead of the approaching midlatitude trough in a low shear zone (Figs 4a and 11a).

To this time, although Hurricane Floyd had undergone a classical hurricane development, it subsequently began to weaken slowly; [48]). As an approaching trough eroded the ridge to its north, Floyd gradually turned northward and propagated parallel to the Florida coast for most of September 15. Around 0630 UTC on September 16 Floyd made landfall near Cape Fear, North Carolina (167 km $h^{-1}$ peak winds) as a strong trough approached from the west (Fig. 11b). A number of tornadoes were reported in North Carolina associated with the storm. Storm surges as high as 9–10 feet (2.75–3.1 m) were reported along the North Carolina coast (Lawrence *et al.* [48]) as Floyd came ashore. Although it was moving rapidly ahead of the approaching trough and remained offshore, rainfall totals exceeding 12 inches (30.5 cm) were recorded in North Carolina, Virginia, Maryland, Delaware, New Jersey, and New York.

Floyd began its transition to an EC as it made landfall, acquiring frontal characteristics, shown in Fig. 4b (Evans and Hart [11]). The evolution of the transitioning Hurricane Floyd as a hybrid cyclone, having characteristics of both tropical and extratropical systems, is apparent in satellite imagery (Fig. 4b and c) and is also well documented in the evolving potential vorticity and thickness analyses of Fig. 11. By 1200 on UTC September 16, ET had commenced: Floyd had begun to interact with the positively tilted midlatitude trough to its northwest (Fig. 11b). The comma-shaped cloud signature associated with Floyd was nearly confined to the north and northwest of the center (Atallah and Bosart [13]), with a weak frontal zone evident offshore (Fig. 4b). Floyd was downgraded to a tropical storm (974 hPa, 60 kt or 111 km $h^{-1}$) at 1800 UTC on September 16. Floyd completed ET, acquiring a cold-core structure, at 1200 UTC on September 17 just prior to making landfall in Canada about 4 h later (Lawrence *et al.* [48]). Upon completion of the ET process, Floyd could be identified as a classic comma pattern in the IR satellite imagery

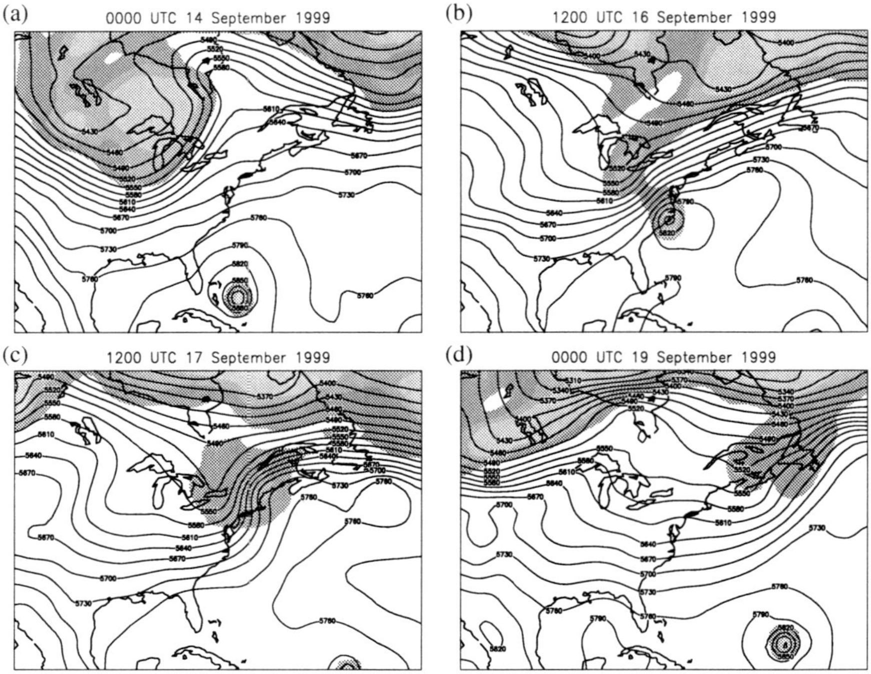


Figure 11: Evolution of Hurricane Floyd (1999), depicted through 325 K isentropic PV (shaded at 1, 2, and 3PVU) and 1000–5000 hPa thickness (contoured every 30 m) analyses from NOGAPS $1° \times 1°$ initialization fields. Times plotted correspond to the satellite imagery in Fig. 4: (a) 0000 UTC September 14 (near peak tropical intensity); (b) 1200 UTC September 16 (ET commenced); (c) 1200 UTC September 17 (ET completed); (d) 0000 UTC September 19 (EC over Canada). Reproduced from Evans and Hart [11].

(Fig. 4c), with the strongest convective activity along a front far to the south of the system center. The PV maximum (>2 PVU) associated with Floyd was located in the baroclinic zone ahead of the weakening approaching trough (Fig. 11c). By 0000 UTC on September 19, Floyd was a weak EC with an intense tropical storm located at the southern tip of the trailing surface front (Fig. 4d). Although it was a baroclinic system when it made landfall in Canada, Floyd was still an intense system with 984 hPa central pressure and 45 kt (83 km $h^{-1}$) peak winds. Thereafter, the cyclone intensified as an EC until September 20, and ultimately merged with another preexisting EC (Evans and Hart [11]).

Throughout its interaction with Floyd, the midlatitude trough underwent a structural evolution reminiscent of the LC1 cycle of Thorncroft *et al.* [50] in which the trough is modulated by a larger-scale anticyclonic environmental shear, as opposed to the cyclonically sheared environment of the LC2 partition. The initially broad, negatively tilted (northwest–southeast) trough evident in Fig. 11a (at 0000 on UTC September 14) became more narrow and acquired a positive tilt by 1200 UTC on September 16 (Fig. 11b) as it interacted with Floyd. By the time of ET completion (1200 UTC on September 17), the trough experienced further weakening and regained a neutral tilt (Fig. 11c) consistent with the southwest–northeast orientation of the jet (Fig. 11c), suggested by Thorncroft *et al.* [50]. With the progressive demise of EC Floyd, the trough continued its poleward contraction as it moved eastward across the North Atlantic (Fig. 11d). If one accepts that this trough evolution is representative of an LC1 life cycle, it can be argued that development of an EC should have been expected in the geographic area where Floyd underwent transition. However, one could also argue that the presence of a large tropical system such as Floyd should be expected to modulate the trough environment, providing an anticyclonic latitudinal shear—or enhancing a preexisting shear zone—ahead of the tropical storm. Therefore, Floyd modified the trough life cycle and the mutual interactions of Floyd and the trough led to both the LC1-type trough evolution and the ET of Floyd. Thus, the chicken and egg cannot be easily distinguished in this case!

## 6 Operational definitions of ET onset and completion

As noted in Section 4, ET occurs in almost all tropical storm basins, yet no universally accepted definition of extratropical transition of TCs presently exists (Malmquist [51]). Forecasters have traditionally used subjective assessments of the effects of a storm's passage over colder waters and the asymmetry of its cloud shield to determine the onset and completion of transition, yet weakening or recurving storms may also exhibit these characteristics (L. Avila 1999, personal communication). Evans and Hart [11] proposed objective diagnostics for ET onset and completion based on the CPS of Hart [34]. Here we contrast these "traditional" and objective approaches to diagnosing ET onset and completion. To do this, we must first review the CPS (Hart [34]) and its relationship to storm structure evolution. We also make comparisons with the satellite-based diagnosis of storm evolution, including the ET process, proposed by Klein *et al.* [22].

TCs are compact weather systems, with relatively symmetric cloud structures and their most intense winds at low levels, close to the storm center. Extratropical transitioning TCs evolve from this structure to become fast moving ($>10\,\mathrm{m\,s^{-1}}$), highly asymmetric systems, in which the peak low-level winds are typically much reduced, whereas the area of gale or storm force winds is often vastly increased. The maximum winds in an EC are found aloft. Thus, to describe the onset and completion of ET comprehensively, measures of both storm symmetry and wind structure must be included—either explicitly or implicitly.

### 6.1 Identification of ET onset

As the ET process begins, a given storm becomes increasingly asymmetric, due to low-level frontogenesis typically resulting from interaction with a midlatitude trough (Evans and Hart [11]; Jones *et al.* [12]; Harr and Elsberry [20]; Klein *et al.* [22]; Foley and Hanstrum [24]). This increase in asymmetry is the key in detecting the onset of transition, but does not guarantee that the system will complete transition—many TCs commence ET, but decay before completing their evolution into a fully cold-core cyclone. In the case of Hurricane Floyd, the comma-shaped cloud signature and associated frontogenesis were obvious signs of a breakdown in storm symmetry, as the storm began transition.

### 6.2 Identification of ET evolution and completion

Tropical storms that remain coherent and persist in becoming increasingly asymmetric are promising candidates for ET. However, some systems in fact do revert to purely tropical structure after they move out of a strongly forced environment. Completion of the ET process signifies that the tropical system has developed midlatitude storm characteristics, in particular the development of a cold-core thermal wind structure (geostrophic wind speed increasing with height), noted by Evans and Hart [11] and Jones *et al.* [12]. In the case of Floyd, it never recovered its structural symmetry; rather, as transition progressed it began to expand in size and continued to increase in asymmetry with peak winds to the right of the track and with a large area of heavy rain to the left of the track. As a further possible condition, Klein *et al.* [22] require that the storm reintensify (in terms of pressure), although classification of ET by the National Hurricane Center [NHC/NWS] is not dependent on this reintensification (L. Avila 1999, personal communication). A storm may continue to decay after recurvature, but its changed structure will be signified by its reclassification as "extratropical". In the context of operational forecasting, it is important to recognize the evolution in the storm structure because it indicates changes in the types and locations of significant weather that one could expect.

### 6.3 Definition of ET onset and completion using the CPS

The synoptic fields and satellite imagery used to infer transition in the previous sections are extremely valuable, but another tool has been developed recently for identifying ET—and other structural changes for the full range of synoptic cyclones. This tool is known as the cyclone phase space, or CPS (Hart [34]). The CPS is a three-parameter summary of fundamental storm structure characteristics necessary to distinguish between TC and EC types. The three parameters are calculated over a 500 km radius region centered on the surface low position (Fig. 12). They are: $B$, the "lower tropospheric" (900–600 hPa) thickness asymmetry; $-V_{\mathrm{T}}^{\mathrm{L}}$, the lower tropospheric (900–600 hPa) thermal wind; and $-V_{\mathrm{T}}^{\mathrm{U}}$, the upper tropospheric (600–300 hPa) thermal wind. Evans and Hart [11] demonstrated the value of this framework for discerning the onset and completion of ET. They defined ET onset as increasing asymmetry, with $B$ exceeding the value of 10 m, which is characteristic of tropical storms and hurricanes. Once the storm has begun to lose symmetry, its lower tropospheric warm core weakens, and it

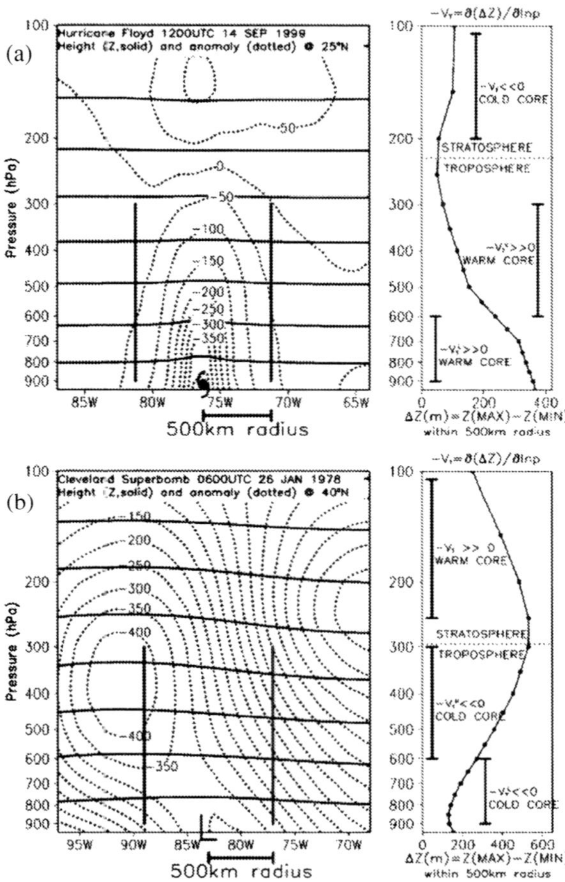


Figure 12: Comparison of radial/height thermal structure in (a) a typical TC, Hurricane Floyd (1999); (b) a typical (strong) EC, the Cleveland "superbomb" of 1978. Note the asymmetry of the EC compared with the TC. The surface low and 500 km radius are indicated. The vertical distribution of the peak height anomaly from the zonal mean at each level (indicative of the thermal wind) is plotted in the right panel in each case Reproduced from Hart [34].

typically transitions to cold core within about 24 h. ET completion is defined as the time at which $-V_{\mathrm{T}}^{\mathrm{L}}$ becomes negative, i.e. the system becomes cold core (Evans and Hart [11]).

To explore the differences in storm evolution for a range of life cycles, we first revisit the case of Hurricane Floyd in terms of the CPS diagnostic. Next, we examine two other cases: decaying tropical storm Hernan (2002) and the ET experience of Bonnie (1998), in terms of CPS, sea surface temperature (SST), and infrared satellite imagery.

### 6.3.1 Hurricane Floyd (1999)

The structural evolution of Hurricane Floyd (1999) depicted using CPS is plotted in Fig. 13: Floyd became an increasingly intense, symmetric tropical system over the week September 8–15. During this time, the symmetry parameter, $B$, remained near 0, while both the lower and upper tropospheric thermal wind parameters became increasingly positive, indicating a strengthening and deepening warm core. Based on the CPS onset definition for ET (Evans and Hart [11]), Floyd commenced ET around 00UTC on September 16 (Fig. 13). Once ET had begun, $-V_T^L$ began to

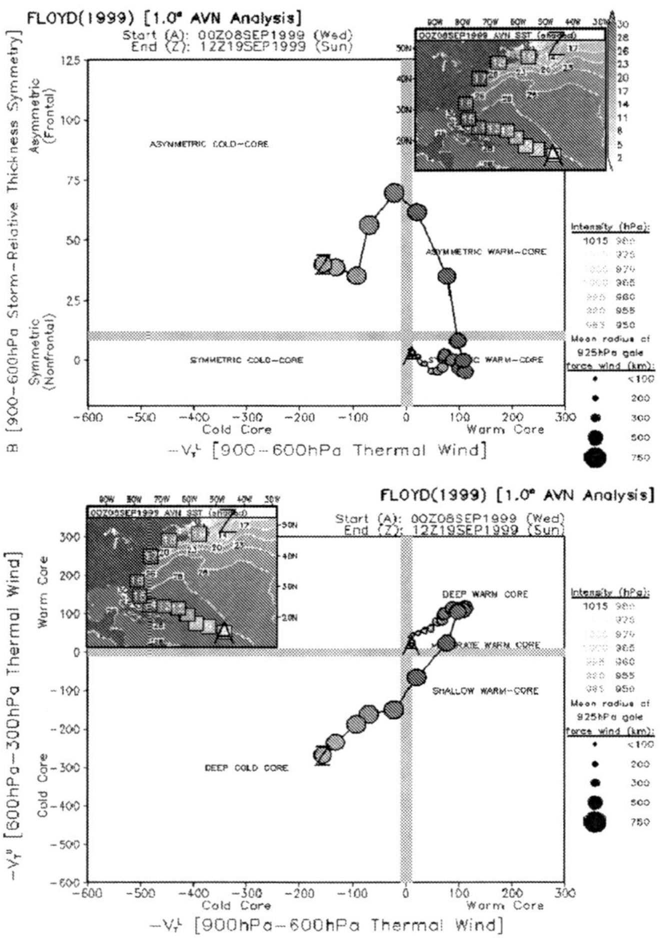


Figure 13: CPS for hurricane Floyd (1999): $V_T^L$ versus $B$ (top); and $V_T^L$ versus $V_T^U$ (bottom). The best track is plotted in the inset of each panel and begins at 0000 UTC September 8 (A), ending at 1200 UTC September 19 (Z). Size (mean 925 hPa gale force wind radius) is represented through dot size. ET commences when $B > 10$ m and completes when $-V_T^L < 0$. Reproduced from Evans and Hart [11].

decrease monotonically, eventually becoming negative at around 00UTC on the 17th and marking the completion of ET (Fig. 13).

### 6.3.2 Tropical storm Hernan (2002) and Hurricane Bonnie (1998)

East Pacific tropical storm Hernan (2002) did not undergo ET; rather it simply decayed. The CPS corresponding to Hernan's evolution, as well as that of Bonnie from 1998 are plotted in Fig. 14. The central pressure of Hernan intensified continually from August 28 through September 2 (Fig. 14b, shading), in agreement

(a)

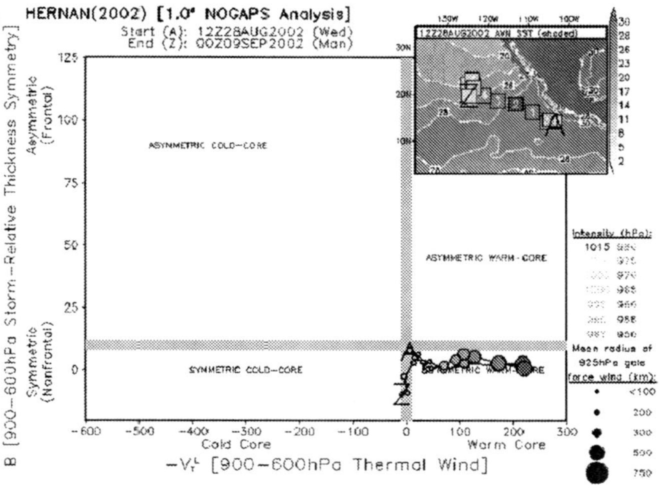


(b)

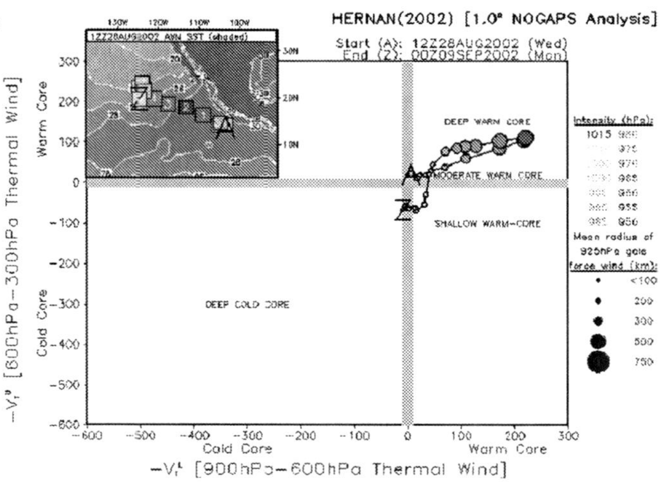


Figure 14: (a) CPS for east Pacific tropical storm Hernan (2002): $V_T^L$ versus $B$; (b) $V_T^L$ versus $V_T^U$. Details of the diagrams are as in Fig. 13. (c) CPS for North Atlantic Hurricane Bonnie (1998): $V_T^L$ versus $B$; and (d) $V_T^L$ versus $V_T^U$.

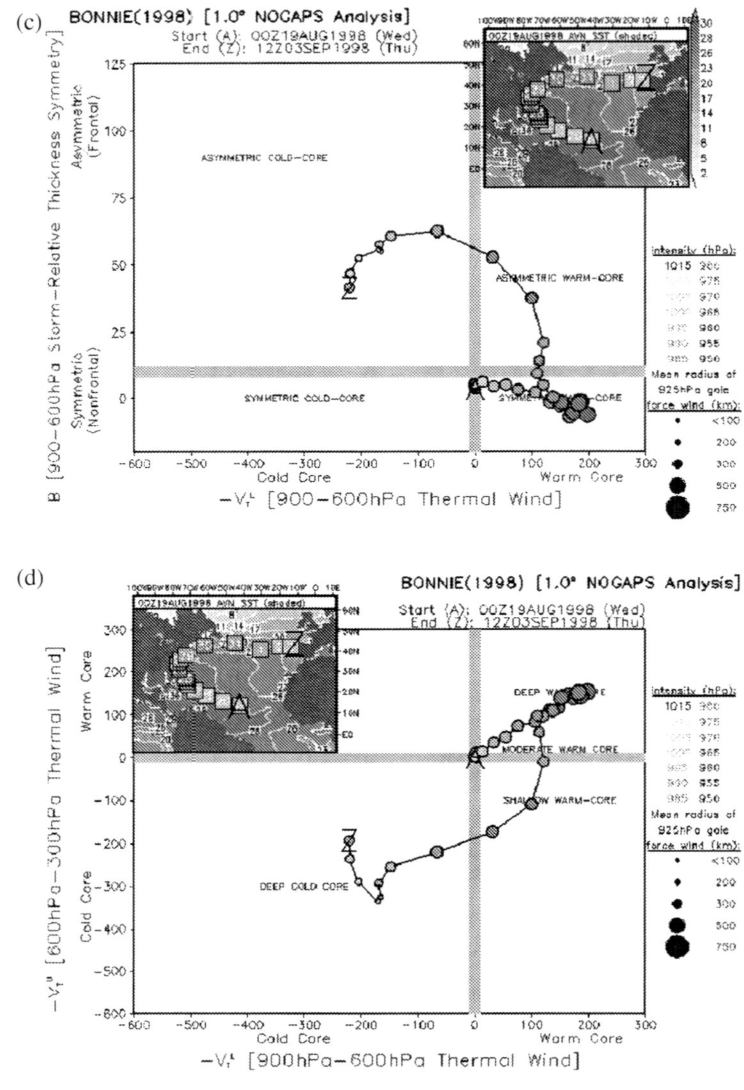


Figure 14 (*continued*)

with the tropical intensification implied by the increasingly positive thermal wind components. Over the subsequent week, Hernan slowly weakened (Fig. 14a and b). This weakening phase coincided with Hernan's movement over cooler SST $< 26°C$ (Fig. 15a) and very weak SST gradients. Drier air is also observed intruding towards the storm center and beginning to erode the eyewall (Fig. 15b). This erosion of the cloud shield to the south was proposed by Klein *et al.* [22] as one identifier of ET, yet this case highlights the need to use this diagnostic with other ET indicators to avoid confusion between the ET process and the purely decaying cases.

(a)

(b)

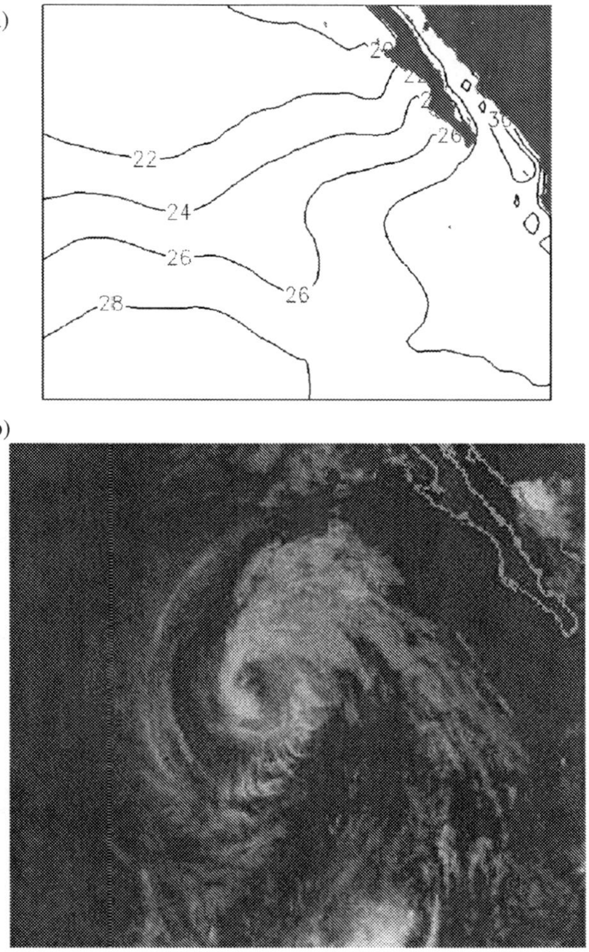


Figure 15: (a) SST and (b) GOES-8 infrared satellite imagery at 1200 UTC September 4, 2002, for decaying east Pacific tropical storm Hernan. (c) SST and (d) GOES-8 infrared satellite imagery for North Atlantic extratropically transitioning Hurricane Bonnie 1200 UTC August 29, 1998.

Both the tropical and post-tropical structural evolution of Hurricane Bonnie (1998) have much more in common with Floyd than they do with Hernan. Bonnie was an intense tropical system that remained coherent through the ET process on August 30, 1998, and continued for many days subsequent to transition, as an EC (Fig. 14c and d). While Bonnie was certainly over waters cooler than 26°C by September 29, she remained near the Gulf Stream (Fig. 15c) and so the SST gradients were much stronger than for the case of Hernan (2002). This may have allowed Bonnie to be impacted by the warmer water source, still relatively close to the south,

(c)

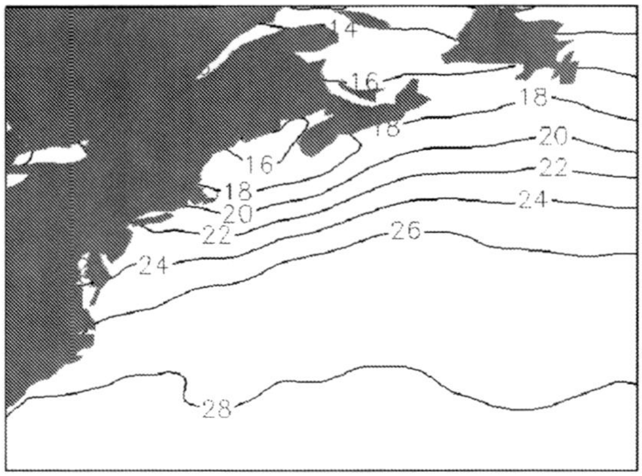


(d)

Figure 15 (*continued*)

and to modify its downstream environment somewhat through continual warm air advection (Fig. 15c). Klein *et al.*'s [22] signature erosion of the cloud shield during ET is evident near the center and to its south (Fig. 15d).

### 6.4 Storm life cycle mapped in terms of the CPS

Arnott *et al.* [14] have taken interpretation of the CPS one additional step. They used K-means cluster analysis to explore the temporal evolution of storm structure (characterized using the CPS) for a set of 19 ET cases. They identified seven distinct clusters representing tropical (three clusters, based on storm strength), extratropical (two clusters), transitioning (one cluster), and hybrid (one cluster) storm structures.

Arnott *et al.* [14] isolated the major path of storm structure evolution through the CPS for these 19 ET cases and calculated storm-centered composites of the synoptic environment for each cluster. The composite environment for cluster #4—the time of ET onset—is very similar to that presented in Fig. 10 (not shown; see Arnott *et al.* [14] for details).

## 7 Summary

Unlike the "conventional" TC, the poleward movement of an extratropically transitioning TC does not simply lead to the decay of the system as it moves into the midlatitudes. TCs undergoing ET typically weaken as they recurve (Evans and McKinley [52]), but as they move into the extratropics these systems can reintensify into intense midlatitude storms with extensive regions of intense rain and larger gale (and even hurricane) force wind areas than their tropical antecedents. The remnant TC can also provide a region of enhanced thermal contrast for the later development of an intense midlatitude storm. The rapid forward speed, the large size of their gale force wind area and intense rainfall region, and the extraordinarily large ocean waves of an ET can persist long after its "tropical storm" status has been discontinued.

ET has been identified to be prevalent across the tropical ocean basins, with the possible exception of the North Indian Ocean. Approximately half of all Atlantic and one-third of west Pacific systems undergo ET. In the Atlantic, almost 50% of these ET systems also make landfall, making ET a real concern for Atlantic hurricane forecasters. Interaction with a synoptic-scale trough is a dominant mechanism for ET, although the tropical system must be sufficiently well supported for tropical development until this time (Hart and Evans [10]).

ET is often poorly forecast by numerical forecast models, providing difficult forecast challenges to the responsible forecast centers around the globe (e.g. Rabier *et al.* [29]). A number of new diagnostics have been proposed to characterize ET: using storms from the western North Pacific, Klein *et al.* [22] have identified satellite imagery signatures and environments typical of ET cases; Evans and Hart [11] have applied the CPS (developed by Hart [34]) as a diagnostic of ET for the North Atlantic. In Atlantic ET cases, the satellite signature proposed by Klein *et al.* [22] is also evident, but it can be found in cases of tropical storms that simply decay (e.g. Hernan 2002). Arnott *et al.* [14] have used cluster analysis to discriminate objectively between different storm structure stages. Using the K-means method, they were able to reproduce the ET onset and completion thresholds arrived at subjectively by Evans and Hart [11]. Composite synoptic environments for ET developed by Arnott *et al.* [14] agree well with those presented here.

## Acknowledgments

This work was supported by the NSF under grant ATM-0351926. The permission of the American Meteorological Society to reproduce Figs 6 and 7 from Hart and

Evans [10], Figs 11 and 16 from Evans and Hart [11], and Fig. 12 from Hart [34] is gratefully acknowledged. We thank Justin Arnott for preparing Fig. 8 and for helpful discussions on the storm stages. We are grateful to Prof. William Gray for permission to reproduce his work in Fig. 2.

## References

[1] Tannehill, I.R., Hurricane of September 16 to 22, 1938. *Monthly Weather Review*, **66**, pp. 286–288, 1938.

[2] Pierce, C.H., The meteorological history of the New England Hurricane of September 21, 1938. *Monthly Weather Review*, **67**, pp. 237–285, 1939.

[3] Palmén, E., Vertical circulation and release of kinetic energy during the development of Hurricane Hazel into an extratropical storm. *Tellus*, **10**, pp. 1–23, 1958.

[4] Kornegay, F.C. & Vincent, D.G., Kinetic energy budget analysis during interaction of tropical storm Candy (1968) with an extratropical frontal system. *Monthly Weather Review*, **104**, pp. 849–859, 1976.

[5] DiMego, G.J. & Bosart, L.F., The transformation of tropical storm Agnes into an extratropical cyclone. Part I: The observed fields and vertical motion computations. *Monthly Weather Review*, **110**, pp. 385–411, 1982.

[6] DiMego, G.J. & Bosart, L.F., The transformation of tropical storm Agnes into an extratropical cyclone. Part II: Moisture, vorticity, and kinetic energy budgets. *Monthly Weather Review*, **110**, pp. 412–433, 1982.

[7] Bosart, L.F. & Bartlo, J.A., Tropical storm formation in a baroclinic environment. *Monthly Weather Review*, **119**, pp. 1979–2013, 1991.

[8] Bosart, L.F. & Lackmann, G.M., Post-landfall tropical cyclone reintensification in a weakly baroclinic environment: a case study of hurricane David (September 1979). *Monthly Weather Review*, **123**, pp. 3268–3291, 1995.

[9] Thorncroft, C. & Jones, S.C., The extratropical transition of hurricanes Felix and Iris in 1995. *Monthly Weather Review*, **128**, pp. 947–972, 2000.

[10] Hart, R.E. & Evans, J.L., A climatology of extratropical transition of Atlantic tropical cyclones. *Journal of Climate*, **14**, pp. 546–564, 2001.

[11] Evans, J.L. & Hart, R.E., Objective indicators of the onset and completion of extratropical transition for Atlantic tropical cyclones. *Monthly Weather Review*, **131**, pp. 909–925, 2003.

[12] Jones, S.C., Harr, P.A., Abraham, J., Bosart, L.F., Bowyer, P.J., Evans, J.L., Hanley, D.E., Hanstrum, B.N., Hart, R.E., Lalaurette, F., Sinclair, M.R., Smith, R.K. & Thorncroft, C., The extratropical transition of tropical cyclones: Forecast challenges, current understanding and future directions. *Weather and Forecasting*, **18**, pp. 1052–1092, 2003.

[13] Atallah, E. & Bosart, L.F., The extratropical transition and precipitation distribution of Hurricane Floyd (1999). *Monthly Weather Review*, **131**, pp. 1063–1081, 2003.

[14] Arnott, J.M., Evans, J.L. & Chiaromonte, F., Characterization of extratropical transition using cluster analysis. *Monthly Weather Review*, **132**, pp. 2916–2937, 2004.

[15] Sekioka, M., A hypothesis on complex of tropical and extratropical cyclones for typhoon in the middle latitudes. I. Synoptic structure of typhoon Marie passing over the Japan Sea. *Journal of the Meteorological Society of Japan*, **34**, pp. 276–287, 1956.

[16] Sekioka, M., A hypothesis on complex of tropical and extratropical cyclones for typhoon in the middle latitudes. II. Synoptic structure of typhoons Louise, Kezia, & Jane passing over the Japan Sea. *Journal of the Meteorological Society of Japan*, **34**, pp. 336–345, 1956.

[17] Sekioka, M., A hypothesis on complex of tropical and extratropical cyclones for typhoon in the middle latitudes. III. Examples of typhoon not accompanied by extratropical cyclone in the middle latitudes. *Journal of the Meteorological Society of Japan*, **35**, pp. 170–173, 1957.

[18] Matano, H. & Sekioka, M., Some aspects of extratropical transformation of a tropical cyclone. *Journal of the Meteorological Society of Japan*, **49**, pp. 736–743, 1971.

[19] Brand, S. & Guard, C.P., Extratropical storm evolution from tropical cyclones in the western North Pacific ocean. *Naval Environmental Prediction Technical Report TR 78-02*, pp. 1–20, 1978.

[20] Harr, P. & Elsberry, R.L., Extratropical transition of tropical cyclones over the western North Pacific. Part I: Evolution of structural characteristics during the transition process. *Monthly Weather Review*, **128**, pp. 2613–2633, 2000.

[21] Harr, P., Elsberry, R.L. & Hogan, T., Extratropical transition of tropical cyclones over the western North Pacific. Part II: The impact of midlatitude circulation characteristics. *Monthly Weather Review*, **128**, pp. 2634–2653, 2000.

[22] Klein, P., Harr, P. & Elsberry, R.L., Extratropical transition of western North Pacific tropical cyclones: an overview and conceptual model of the transformation stage. *Weather and Forecasting*, **15**, pp. 373–396, 2000.

[23] Sinclair, M.R., Extratropical transition of southwest Pacific tropical cyclones. Part I: Climatology and mean structure changes. *Monthly Weather Review*, **130,** pp. 590–609, 2002.

[24] Foley, G.R. & Hanstrum, B.N., The capture of tropical cyclones by cold fronts off the west coast of Australia. *Weather and Forecasting*, **9**, pp. 577–592, 1994.

[25] Prater-Mayes, B. & Evans, J.L., Sensitivity of modeled tropical cyclone track and structure of Hurricane Irene (1999) to the convection parameterization scheme. *Meteorology and Atmospheric Physics*, **80**, pp. 103–115, 2002.

[26] Evans, J.L. & Prater-Mayes, B., Factors affecting the post-transition intensification of Hurricane Irene (1999). *Monthly Weather Review*, **132**, pp. 1355–1368, 2004.

[27] Agusti-Panareda, A., Thorncroft, C.D., Craig, G.C. & Gray, S.L., The extratropical transition of hurricane Irene (1999): a potential vorticity perspective. *Quarterly Journal of the Royal Meteorological Society*, **130**, pp. 1–28, 2004.

[28] Abraham, J.D., *et al.*, Extratropical transition. *Fifth WMO/ICSU Int. Workshop on Tropical Cyclones (IWTC- V), Topic 4.6*, Cairns, Australia, December 1–13, 2002.

[29] Rabier, F., Klinker, E., Courtier, P. & Hollingsworth, A., Sensitivity of forecast errors to initial conditions. *Quarterly Journal of the Royal Meteorological Society*, **122,** pp. 121–150, 1996.

[30] Bowyer, P.J., Phenomenal waves with a transitioning tropical cyclone (Luis, the Queen, and the buoys). *Preprints, 24th AMS Conference on Hurricanes and Tropical Meteorology*, Fort Lauderdale, FL, pp. 294–295, 2000.

[31] Moon, I.-J., Ginis, I., Hara, T., Tolman, H., Wright, C.W. & Walsh, E.J., Numerical simulation of sea-surface directional wave spectra under hurricane wind forcing. *Journal of Physical Oceanography*, **33**, pp. 1680–1706, 2003.

[32] MacAfee, A.W. & Bowyer, P.J., Trapped-fetch waves in a transitioning tropical cyclone (Part I—The need and the theory). *Preprints, 24th AMS Conference on Hurricanes and Tropical Meteorology*, Fort Lauderdale, FL, pp. 292–293, 2000.

[33] MacAfee, A.W. & Bowyer, P.J., Trapped-fetch waves in a transitioning tropical cyclone (Part II—Analytical and predictive model). *Preprints, 24th AMS Conference on Hurricanes and Tropical Meteorology*, Fort Lauderdale, FL, pp. 165–166, 2000.

[34] Hart, R.E., A cyclone phase space derived from thermal wind and thermal asymmetry. *Monthly Weather Review*, **131**, pp. 585–616, 2003.

[35] Hatada, T., *The Mongol Invasions* (in Japanese), Chuokoron Sha: Tokyo, 1965.

[36] Yanai, M., *English translation of the letter by the Khubilai Khan to the Japanese emperor in 1266*, Meteo 216B class notes. Available from Prof. Yanai, UCLA, 1975.

[37] Gray, W.M., Hurricanes: their formation, structure and likely role in the tropical circulation. *Meteorology Over the Tropical Oceans*, ed. D.B. Shaw, *Royal Meteorological Society*, pp. 155–218, 1979.

[38] Emanuel, K.A., Sensitivity of tropical cyclones to surface exchange coefficients and a revised steady-state model incorporating eye dynamics. *Journal of the Atmospheric Sciences*, **52**, pp. 3969–3976, 1995.

[39] Kalnay, E., Kanamitsu, M., Kistler, R., Collins, W., Deaven, D., Gandin, L., Iredell, M., Saha, S., White, G., Woollen, J., Zhu, Y., Leetmaa, A., Reynolds, B., Chelliah, M., Ebisuzaki, W., Higgins, W., Janowiak, J., Mo, K.C., Ropelewski, C., Wang, J., Jenne, R. & Joseph, D., The NCEP/NCAR 40-year reanalysis project. *Bulletin of the American Meteorological Society*, **77**, pp. 437–472, 1996.

[40] Neumann, C.J., Jarvinen, B.R., McAdie, C.J. & Elms, J.D., *Tropical Cyclones of the North Atlantic Ocean 1871–1992*. NOAA Historical Climatology Series 6–2, National Climatic Data Center. Asheville, NC, 1993.

[41] Gray, W.M., Global view of the origin of tropical disturbances and storms. *Monthly Weather Review*, **96**, pp. 669–700, 1968.

[42] Gray, W.M., Tropical cyclone genesis and intensification. *Intense Atmospheric Vortices*, eds L. Bengtsson & J. Lighthill, Springer: New York, pp. 3–20, 1982.

[43] Molinari, J., Knight, D., Dickinson, M., Vollaro, D. & Skubis, S., Potential vorticity, easterly waves and eastern Pacific tropical cyclogenesis. *Monthly Weather Review*, **125**, pp. 2699–2708, 1997.

[44] Dickinson, M. & Molinari, J., Mixed Rossby-gravity waves and western Pacific tropical cyclogenesis. Part I: Synoptic evolution. *Journal of the Atmospheric Sciences*, **59**, pp. 2183–2196, 2002.

[45] Bister, M. & Emanuel, K.A., The genesis of Hurricane Guillermo: TEXMEX analyses and a modeling study. *Monthly Weather Review*, **124**, pp. 2662–2682, 1997.

[46] Evans, J.L. & Jaskiewicz, F.A., Satellite-based monitoring of intraseasonal variations in tropical Pacific and Atlantic convection. *Geophysical Research Letters*, **28**, pp. 1511–1514, 2001.

[47] Grandau, F. & Engel, G., *JTWC 2001 Annual Tropical Cyclone Report*. Available from https://metoc.npmoc.navy.mil/jtwc/atcr/2001atcr, 2001.

[48] Lawrence, M.B., Avila, L.A., Beven, J.L., Franklin, J.L., Guiney, J.L. & Pasch, R.J., Atlantic hurricane season of 1999. *Monthly Weather Review*, **129**, pp. 3057–3084, 2001.

[49] Pasch, R.J., Kimberlain, T. & Stewart, S.R., Preliminary report on Hurricane Floyd September 7–17, 1999. National Hurricane Center, 1999 [Available from the Tropical Prediction Center, Miami, FL; also available online at http://www.nhc.noaa.gov/1999/floyd.html].

[50] Thorncroft, C.D., Hoskins, B.J. & McIntyre, M.E., Two paradigms of baroclinic wave life cycle behavior. *Quarterly Journal of the Royal Meteorological Society*, **119**, pp. 17–55, 1993.

[51] Malmquist, D. Meteorologists and insurers explore extratropical transition of tropical cyclones. *EOS*, February 16, 1999.

[52] Evans, J.L. & McKinley, K., Relative timing of tropical storm lifetime maximum intensity and track recurvature. *Meteorology and Atmospheric Physics*, **65**, pp. 241–245, 1998.

# CHAPTER 8

# Historical and possible future changes of wave heights in northern hemisphere oceans

X.L. Wang & V.R. Swail
*Climate Research Branch, Meteorological Service of Canada, Downsview, Ontario, Canada.*

## Abstract

In this study, we review and extend previous studies on changes of wave heights, focusing on both observed and possible future changes in the North Atlantic and in the North Pacific. Two wave hindcasts are used here as observed waves for 1958–1997: a global wave hindcast based on the 10-m winds taken from the NRA (i.e., the reanalysis of the National Centers for Environment Prediction, National Center for Atmospheric Research), and a detailed North Atlantic wave hindcast produced with intensively reanalyzed surface winds over the North Atlantic basin.

The observed relationships between sea level pressure (SLP) and significant wave height (SWH) are used to construct climate change scenarios of SWH.

Scenarios of seasonal mean SWH are constructed by means of redundancy analysis (RA), while projections of seasonal extreme SWH are made using nonstationary generalized extreme value (GEV) models. The RA and GEV models are trained using the NRA SLP and the observed SWH data for 1958–1997. The SWH scenarios are constructed using seasonal mean SLP and squared SLP gradient fields from a coupled climate model (CGCM2) under three different forcing scenarios. The projected (and observed) trends/changes in SWH are assessed by conducting a trend analysis, in which linear trends are evaluated against quadratic trends.

Both oceans had significant changes in both winter and fall seasonal means and extremes of SWH during 1958–1997; they are also projected to have significant changes in the 21st century under all three forcing scenarios. The rate and sign of the projected future SWH changes are not constant throughout the 21st century; and in some regions, these appear to be quite dependant on the forcing conditions. Often, the projected SWH changes are characterized either by faster increases in the late decades than in the early decades or by decreases in the early decades followed by increases, depending on the forcing scenario and the specific location. The rate of SWH change appears to have a positive relationship with the rate of increase in the greenhouse gases forcing.

# 1 Introduction

The constantly changing climate of the Earth consists of both natural climate variability and human-induced climate change. Global mean temperature is projected to rise by 1.4–5.8 K between 1990 and 2100 (IPCC [1]) for the full range of "SRES" (Special Report on Emissions Scenarios) forcing scenarios; the corresponding rise in sea level is projected to be 0.09–0.88 m (IPCC [2]).

The oceans are an important component of the climate system and ocean borne commerce is sensitive to the state of the ocean surface. Amongst other surface characteristics, ocean wave heights could be affected by anthropogenic forcing of the climate system. Information on ocean wave climate and its variability and trends are required for a wide range of applications, including the determination of engineering design parameters for offshore oil and gas exploration, production and transportation, coastal development, and the detailed planning of shipping routes. Since the design of marine and coastal infrastructure is constrained by the largest wave height event anticipated during a fixed design period, increases in the extremes of wave heights could have an impact on the life span of these installations that will be in excess of impacts anticipated from the rising sea level.

Analysis of the observed variability and trends of ocean wave heights, and projections of its possible future changes, are therefore useful for the design and operation of coastal and offshore industries. This study aims to review and extend the previous studies related to trends and variability of ocean wave heights, focusing on both observed and possible future changes in the North Atlantic and the North Pacific.

The remainder of this paper is organized as follows: The datasets and methodologies used in this study are briefly described in Section 2. Previous studies on historical changes of wave heights, as derived from *in situ* observations and from numerical wave hindcasts, are reviewed and extended in Section 3. Possible future changes in seasonal mean and extreme wave heights are discussed in Section 4. We complete this paper with some concluding remarks in Section 5.

# 2 Datasets and methodologies

## 2.1 Datasets

Our previous studies (Wang and Swail [3, 4]) revealed that, on the seasonal time scale, there is a close relationship between significant wave height (SWH) and sea level pressure (SLP) in the data observed during 1958–1997, especially in the cold seasons (winter and fall). In this study, winter is defined as January–February–March (JFM), and fall, October–November–December (OND). The observed

SLP–SWH relationship in each season is represented by a pair of regression models: redundancy analysis for seasonal mean SWH (see Section 2.2), and nonstationary generalized extreme value (GEV) analysis for seasonal maxima of SWH (see Section 2.3). Assuming that the observed SLP–SWH relationship will continue to hold under the possible future climates as projected by a coupled global climate model, we then used this relationship to project possible future changes of SWH in the northern hemisphere oceans, using climate model projections of possible future SLP changes as predictor. Therefore, observed SWH and SLP data are needed to assess historical changes and train the regression models, and projections of possible future SLP changes are needed to make projections of possible future SWH changes.

Specifically, the following four datasets were involved in this study:

- Observed SLP data. Seasonal means of SLP for the period from 1958 to 1997 were derived from twice-daily data of the NRA (the reanalysis of the National Centers for Environment Prediction, National Center for Atmospheric Research; Kalnay *et al.* [5]). These data had previously been converted to the global 96 × 48 Gaussian grid (96 × 48 grid points, spacing approximately 3.75° latitude by 3.7° longitude) used by CGCM2 (the second generation coupled GCM of the Canadian Centre for Climate Modelling and Analysis; Flato and Boer [6]).
- Observed Atlantic SWH data. Seasonal means and maxima of SWH over the North Atlantic (80°W–20°E, 20°N–70°N) were derived from 6-hourly wave data of a detailed North Atlantic wave hindcast. The hindcast was obtained by driving a third generation ocean wave model (OWI 3G) with surface winds over the North Atlantic basin that were intensively reanalyzed (for more details see Wang and Swail [3]; Swail and Cox [7]). The wave data are on a 0.625° × 0.833° latitude/longitude grid, covering the 40-year period from 1958 to 1997.
- Observed Pacific SWH data. Seasonal means and maxima of SWH over the North Pacific (120°E–270°E, 20°N–70°N) were derived from 6-hourly wave data of a global wave hindcast (Cox and Swail [8]), which was produced by driving the ODGP2 wave model (second generation formation of the original Ocean Data Gathering Program wave model; see Cardone *et al.* [9]) with the NRA 6-hourly 10-m wind fields. The wave data are on a 1.25° × 2.5° latitude/longitude grid over the globe, for the 40-year period from 1958 to 1997.
- Climate model projected SLP data. Projections of future seasonal mean SLP were obtained from three ensembles of CGCM2 simulations using three different forcing scenarios: (i) a modified version of the IPCC IS92a scenario, in which the change in greenhouse gases (GHG) forcing corresponds to that observed from 1850 to 1990 and increases at a rate of 1% per year thereafter until the year 2100 (cf. Boer *et al.* [10]); (ii) the SRES A2 forcing scenario, which is described in detail in the Special Report on Emissions Scenarios (IPCC [2]); (iii) the SRES B2 forcing scenario. The forcing includes both GHG and aerosol loadings. The A2 scenario is similar to the IS92a scenario. The B2 scenario reflects slower economic development and population growth and thus a slower increase in GHG forcing. For each forcing scenario, an ensemble of three integrations was carried out with different initial conditions. The three integrations

in an ensemble can be considered as three independent realizations of the same stochastic process. The integration period is 251 years (1850–2100) for the IS92a scenario and 111 years (1990–2100) for both the A2 and B2 scenarios.

Both observed and projected SLP data are on the global 96×48 Gaussian grid, but only those data over the North Atlantic sector (82.5°W–22.5°E, 16.7°N–72.36°N) and the North Pacific sector (116.25°E–86.25°W, 16.7°N–72.36°N) were used in this study. Monthly means of SLP were also used to calculate monthly, as well as seasonal mean, fields of squared SLP gradient. Then, the seasonal mean squared SLP gradients were expressed as anomalies relative to their 1961–1990 baseline climates (the CGCM2 simulated baseline climate was calculated from the IS92a scenario ensemble simulations; CGCM2 simulated SLP gradients were obtained from the simulated monthly means of SLP after they had been adjusted to the observed baseline climate). Actually, all seasonal quantities of SLP and SWH were re-expressed as anomalies relative to the related baseline climate. These anomalies were then used as the predictor and predictand data in the statistical "downscaling" models (see Sections 2.2 and 2.3).

## 2.2 Redundancy analysis

For the seasonal means, the observed SLP–SWH relationship was determined by using a least squares regression technique called redundancy analysis (RA), which seeks to maximize the associated predictand variance while searching for a hierarchy of the best predicted components and the corresponding predictor components. Details about RA can be found in Wang *et al.* [11], Wang and Swail [3, 4], Wang and Zwiers [12], von Storch and Zwiers [13], and Tyler [14].

RA was performed in reduced dimension versions of the predictor and predictand datasets so as to focus attention on their large-scale behavior. Thus the seven leading principal components (PCs) of anomalies of seasonal mean SWH and seasonal mean SLP (over the North Atlantic and the North Pacific, separately) were used instead of the original datasets in the RA-regression analysis, which are associated with 75–90% of the total variance for the SWH, and 81–94% for the NRA SLP (see Table 1). The CGCM2 simulated SLP anomalies were projected onto the retained EOFs (empirical orthogonal functions) derived from NRA SLP anomalies to obtain seven simulated pseudo-PCs of SLP anomalies (see Wang *et al.* [11] for the rationale behind this). These seven pseudo-PCs represent 87–89% (82–85%) of the total simulated variance of winter (fall) SLP anomalies in the North Atlantic, and 75–80% (65–70%) in the North Pacific. Most importantly, the observed changes in SWH and SLP are well represented in the retained leading PCs (Wang and Swail [3]).

CGCM2-simulated anomalies of seasonal mean SLP were then substituted into the RA regression model to project future anomalies of seasonal mean SWH of reduced dimension. These EOF–space projections (i.e., projected PCs) were then converted back to the physical space using the related EOFs of SWH anomalies. The projected SWH anomalies obtained in this way were superimposed on the observed

Table 1: Redundancy index ($R^2$) of the RA model for seasonal mean SWH (also denoted as Havg) over the North Atlantic and the North Pacific, as well as the correlation coefficients between the observed and RA-predicted leading PCs of Havg for winter (JFM) and fall (OND). Also shown are the percentages of the total variance in the mean SWH ($P_{SWH}$) or in the SLP fields ($P_{SLP}$) that are associated with the seven retained EOFs for each season.

| Sector | Season | $P_{SLP}$ | $P_{SWH}$ | $R^2$ | PC1 | PC2 | PC3 |
|---|---|---|---|---|---|---|---|
| North Atlantic | JFM | 94.0 | 89.3 | 0.96 | 0.96 | 0.79 | 0.55 |
| | OND | 90.9 | 83.3 | 0.81 | 0.79 | 0.90 | 0.75 |
| North Pacific | JFM | 88.0 | 84.5 | 0.85 | 0.93 | 0.90 | 0.62 |
| | OND | 81.5 | 74.8 | 0.72 | 0.85 | 0.62 | 0.30 |

SWH baseline climate to obtain projections of the full seasonal mean SWH field. A separate projection was produced using the SLP anomalies derived from each member of each three-member ensemble of CGCM2 simulations, resulting in an ensemble of three projections of seasonal mean SWH for each of the three forcing scenarios. These projections of seasonal mean SWH were then analyzed to assess projected trends in ocean wave heights (see Sections 2.4 and 4). More details about using RA to project possible future anomalies of seasonal mean SWH were described in Wang *et al.* [11].

### 2.3 GEV analysis

The approach used for making projections of future SWH extremes in this study is the same as detailed in Wang *et al.* [11], which is based on a GEV distribution GEV($\mu$, $\sigma$, $\xi$) where the location and scale parameters ($\mu$ and $\sigma$) are allowed to vary with time-dependant covariates. Specifically, the following five nested GEV models were fitted to observed seasonal maxima of SWH at each grid point (derived from the related wave hindcast; see Section 2.1):

$$\mathrm{GEV}_0(\mu, \sigma, \xi), \tag{1}$$

$$\mathrm{GEV}_1(\mu_t = \mu_0 + \rho_1 P_t, \sigma, \xi), \tag{2}$$

$$\mathrm{GEV}_2(\mu_t = \mu_0 + \rho_1 P_t + \rho_2 G_t, \sigma, \xi), \tag{3}$$

$$\mathrm{GEV}_3(\mu_t = \mu_0 + \rho_1 P_t + \rho_2 G_t, \log(\sigma_t) = \lambda_0 + \lambda_1 P_t, \xi), \tag{4}$$

$$\mathrm{GEV}_4(\mu_t = \mu_0 + \rho_1 P_t + \rho_2 G_t, \log(\sigma_t) = \lambda_0 + \lambda_1 P_t + \lambda_2 G_t, \xi). \tag{5}$$

The time-dependent covariates used in these models are the seasonal mean SLP anomaly ($P_t$) and the anomaly of squared SLP gradient ($G_t$) at the SWH grid point. The above GEV models were trained using observed (NRA) $P_t$ and $G_t$ for the 40-year period from 1958 to 1997. Note that the SLP quantities that are required on the fine SWH grid were calculated from values at four nearest SLP grid points, using weights proportional to the inverse of the distance.

The significance of the linear relationships built into the various GEV models (and the goodness of fit of the GEV models themselves) was assessed by performing likelihood ratio tests (see Wang *et al.* [11] for more details). Results of the tests show that the location parameter for the SWH extremes is significantly correlated with both the seasonal mean SLP anomaly ($P_t$) and the anomaly of squared SLP gradient ($G_t$), and that the scale parameter appears to be independent of either $P_t$ or $G_t$ (see Table 2 and Fig. 1; also figure 1 of Wang *et al.* [11]). In other words, the $GEV_2$ model is the model of best fit and hence was used to project possible future changes in SWH extremes. To this end, the CGCM2-simulated values of $P_t$ and $G_t$ were substituted into the fitted expression for the location parameter to produce a time series of location parameter estimates for the period 1990–2099 (110 years):

$$\hat{\mu}_t = \hat{\mu}_0 + \hat{\rho}_1 P_t + \hat{\rho}_2 G_t \quad (t = 1, 2, \ldots, 110). \tag{6}$$

(In this study, $\hat{y}$ denotes an estimate of $y$.) This time series was then subject to a trend analysis as outlined in Section 2.4. The trend component of the location parameter, in turn, was used to project changes in the size or frequency of extreme SWH events (see Section 4).

## 2.4 Trend analysis

Trends in the time series of the observed/projected seasonal mean SWH ($h_t$) and the projected GEV location parameter at each grid point were analyzed using a regression technique.

Table 2: The field significance of the regression parameters $\rho_1$, $\rho_2$, $\lambda_1$, and $\lambda_2$, as represented by the percentages of grid points where a GEV model is favored over a relatively simpler GEV model at the 5% level of significance (Wang *et al.* [11]).

| GEV models evaluated | North Atlantic | | North Pacific | |
|---|---|---|---|---|
| | JFM | OND | JFM | OND |
| $GEV_1$ over $GEV_0$ ($\rho_1$) | 54.0 | 44.8 | 37.8 | 38.4 |
| $GEV_2$ over $GEV_1$ ($\rho_2$) | 43.6 | 29.7 | 33.5 | 27.4 |
| $GEV_2$ over $GEV_0$ ($\rho_1$ and $\rho_2$) | 70.3 | 53.1 | 50.4 | 49.1 |
| $GEV_3$ over $GEV_2$ ($\lambda_1$) | 2.3 | 2.2 | 7.4 | 7.5 |
| $GEV_3$ over $GEV_0$ ($\rho_1$, $\rho_2$, and $\lambda_1$) | 31.2 | 21.4 | 45.8 | 44.9 |
| $GEV_4$ over $GEV_3$ ($\lambda_2$) | 1.4 | 1.0 | 2.2 | 3.6 |
| $GEV_4$ over $GEV_0$ ($\rho_1$, $\rho_2$, $\lambda_1$, and $\lambda_2$) | 10.4 | 8.4 | 12.8 | 15.0 |

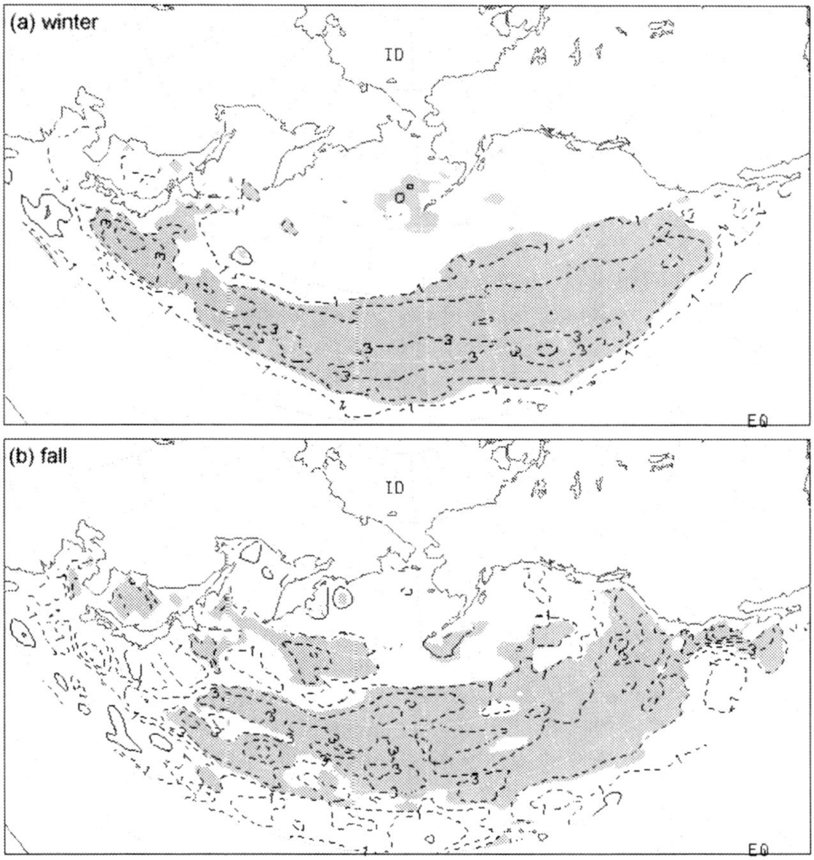


Figure 1: The regression coefficient between the location parameter of (a) winter and (b) fall seasonal maximal SWH and anomalies of seasonal mean SLP ($\rho_1$) over the North Pacific. The contour interval is 1.0. Shading indicates locations where the regression coefficient is significantly (at the 5% level) different from zero.

The following regression models (RMs) were fitted to the time series $x_t$ (where $x_t = h_t$ or $x_t = \hat{\mu}_t$) at each grid point:

$$\text{RM}_0\text{: } x_t = \alpha_0 + \varepsilon_t \text{ (i.e., no trends in } x_t\text{)}, \tag{7}$$

$$\text{RM}_1\text{: } x_t = \alpha_0 + \alpha_1 t + \varepsilon_t \text{ (i.e., a linear trend in } x_t\text{)}, \tag{8}$$

$$\text{RM}_2\text{: } x_t = \alpha_0 + \alpha_1 t + \alpha_2 t^2 + \varepsilon_t \text{ (i.e., a quadratic trend in } x_t\text{)}, \tag{9}$$

where $\alpha_i$ ($i$ = 0, 1, 2) are the regression parameters and $\varepsilon_t$ denotes a zero-mean white noise process. The relative validity of these models is assessed by performing $F$-tests (see Wang *et al.* [11] for more details).

The sample that we used to perform the trend analysis for each forcing scenario was obtained by combining the three members of each ensemble into a single sample with three "observations" at each sampling time. Thus, the series length is $n = 3 \times 110$ for the projected time series. Results of the $F$-tests are discussed in detail in Section 4.

## 3 Historical changes of wave heights

In the early 1990s, Bacon and Carter [15, 16] raised the issue of wave climate changes in the North Atlantic and North Sea by presenting a review of all available data describing long-term trends in the wave climate, using both visual estimates and instrumental measurements of wave heights. They found that there exists an increase in mean wave height over the whole of the North Atlantic, possibly since 1950, of about 2% per year, and that the long-term changes in wave climate are linked to the atmospheric pressure gradient measured between the Iceland low and the Azores high. Bouws *et al.* [17] also reported a similar rate of increase (2–3 cm/year) in mean wave heights observed in the North Atlantic. However, the rate of wave height increase is very likely overestimated in these studies, because their data from wave charts for the 1960s and 1970s very likely underestimated the wave heights for that period (Wang and Swail [3]).

In recent years there have been several attempts to investigate the variability of waves and storms in the North Atlantic Ocean. The first of these studies was carried out by Kushnir *et al.* [18], who generated a statistical hindcast to quantitatively estimate the systematic increase in wave heights over the period 1962–1986. They found that an increasing trend in SWH at several northeast Atlantic locations since 1960 or so was related to the systematic deepening of the Icelandic low and intensification of the Azores high over the last three decades. Their analysis also suggested that wave heights south of 40°N have decreased during the same period. Subsequently, a major effort was undertaken by the European Union Waves and Storms in the Atlantic (WASA) project (WASA [19]; Gunther *et al.* [20]), which produced comprehensive estimates of linear trends in means and extremes of annual significant wave heights. Unfortunately, inhomogeneities and other problems with the hindcast in the area west of 20°W limited the reliability of the analysis to the extreme eastern North Atlantic, where it was concluded that the wave climate has undergone significant variations on a time scale of decades. Part of that variability was found to be related to the North Atlantic Oscillation (NAO). In spite of an apparent artificial worsening of the storm climate in data-sparse areas (Gunther *et al.* [20]), the WASA hindcast produced a well-validated reconstruction of the wave climate of the northeast Atlantic based on the best information and technology available at that time.

While much attention has been drawn to the characterization of trends and variability of waves in the North Atlantic, only a few studies have attempted to assess trends and variability of waves in the North Pacific. The first of these is by Allan and Komar [21], who analyzed wave measurements obtained from six deep-water buoys extending from the Gulf of Alaska to southern California. They found that wave heights have increased during the past 20–30 years. The greatest increase (4.2 cm/year for winter mean SWH, and 2.7 cm/year for annual mean SWH) has occurred offshore from the Washington coast, with a systematic reduction in the rate of SWH increase southward along the west coast. There is an increase of 2.1–3.1 cm/year for winter mean SWH offshore from Oregon and northern California, while no significant increase has occurred off the coast of southern–central California or in the Gulf of Alaska. They reported that the 100-year return value of SWH estimated from wave data up through 1996 (about 10 m) was exceeded six times during 1997–1999, and that the rise in wave heights had been associated with an increase in the wave period that has little systematic trend variation along the west coast. Allan and Komar [21] also found that the interannual variability of the waves above and below the long-term trends is correlated with the Southern Oscillation and also with the East Pacific pattern, a measure of the strength of the westerly winds and the position of the jet stream, both of which are important to the strength of storms and wave generation.

The advent of long-term climate reanalysis projects undertaken by major numerical weather prediction centers significantly changed the approach to the investigation of climate variability, and to a lesser extent, trend analysis on decadal and longer time scales. Inhomogeneities associated with improvements in data analysis techniques, and evolution and upgrades in operational numerical models have led these centers to major attempts to produce a consistent analysis of the atmosphere through so-called "reanalysis" using historical atmospheric observations and current analysis schemes and numerical weather prediction models. Reanalysis does not account for changes in time of data sources, density or quality, thus leaving some unknown degree of temporal inhomogeneity. Two such projects have been widely used for a variety of climate research applications: (i) a preliminary global atmospheric reanalysis produced at the European Centre for Medium-Range Weather Forecasts (ECMWF) for the 15-year period from 1979 to 1993 (ERA-15, Gibson *et al.* [22]); (ii) the NRA Project (cf. Section 2.1). In the past 3 years, the ECMWF has been conducting ERA-40, a reanalysis of global atmospheric wind, temperature and humidity fields, stratospheric ozone, deep water sea states, and soil conditions for the 45 year period from 1957 to 2002, using a coupled atmosphere–wave model with variational data assimilation. The ERA-40 will produce the most complete wave dataset on a $1.5° \times 1.5°$ latitude/longitude grid covering the whole globe.

The first wave–climate study using reanalysis data was undertaken by Sterl *et al.* [23], who analyzed the linear trend of the monthly mean SWH for each calendar month over the 15-year ERA period. Large trends were found in January in the North Atlantic mean SWH, and in July in the ocean south of Africa; the trends were found to vary greatly from month to month. While the Sterl *et al.* [23] hindcast represented a major advance in wind field quality and homogeneity, the time period was too short to confirm a significant change in the global wave climate.

The first 40-year wave hindcast based on reanalysis products was a global hindcast carried out by Environment Canada using the NRA 6-hourly 10-m winds (Cox and Swail [8]; see also Section 2.1), which was found to have very good agreement with observations up to the 99 percentiles of SWH. In one of our previous studies (Wang and Swail [4]), we analyzed this global wave hindcast for trends and variability in seasonal extremes (90 and 99 percentiles) of SWH. We first used the Mann–Kendall test to assess linear trends of SWH extremes on seasonal time scales, accounting for the effect of autocorrelation when assessing trends and their significance level. We then used redundancy analysis to characterize the links between the large-scale variability of SWH extremes and large-scale atmospheric circulation (SLP) patterns, which were further used to extend the numerical hindcast back to January 1899, providing a best guess of the historical variability of SWH extremes. As a result, we found that increases of SWH in the northeast Atlantic are particularly significant in the winter season for the last four decades, which are matched by significant decreases in the subtropical North Atlantic. These trends were found to be closely connected to significant changes in the variability of the NAO. For the North Pacific, we found that SWH changes featuring increases (up to 3.1 cm/year) in the central North Pacific are more significant in winter and spring (April–May–June) and that these changes were connected with a deeper and more eastward extended Aleutian low. From a 50-year wave hindcast covering December–March 1948/49 through 1997/98 (using the NRA 6-hourly 10-m winds), Graham *et al.* [24] and Graham and Diaz [25] also identified a similar pattern of wave height changes in the NP. However, based on the statistical hindcast, Wang and Swail [4] found no significant trends in SWH extremes for the whole last century although long-term variability (e.g., multidecadal fluctuation) is especially noticeable in this basin.

In parallel with the initial NRA-based global wave hindcast, a detailed high-resolution wave hindcast was produced for the North Atlantic (Swail and Cox [7]). The objective of that project was to produce an engineering-quality, homogeneous, long-term wind and wave database for assessment of the wave climate of the North Atlantic, its trends, and variability. The most important feature of the hindcast was the rigorous attention devoted to producing the wind fields used to drive the wave model. To remove potential biases in the historical wind fields, all wind observations from ships and buoys were reassimilated into the analysis taking account the method of observation, anemometer height, and stability. Wind fields for all significant storms were then painstakingly kinematically reanalyzed. Furthermore, high-resolution surface wind fields for all tropical cyclones, as specified by a proven tropical cyclone boundary layer model, were assimilated into the wind fields to provide greater skill and resolution in the resulting wave hindcasts. Finally, these high-quality wind fields were used to force a much higher resolution, third generation wave model. The resulting wave hindcast (i.e., the North Atlantic hindcast; see also Section 2.1) has been shown to make significant improvement in resolving the most extreme storms, including tropical storms (Swail and Cox [7]). Caires *et al.* [26] compared this hindcast with several other reanalyses (including the latest reanalysis ERA-40) and concluded that this North Atlantic hindcast best represents the measurements within that basin.

In one of our recent studies (Wang and Swail [3]) we compared changes of SWH derived from the North Atlantic hindcast with those derived from *in situ* observations and from previous wave hindcasts (Carter and Draper [27]; Bacon and Carter [16]; Bouws *et al.* [17]; Gunther *et al.* [20]). We also carried out a detailed analysis for trends in extremes (90 and 99 percentiles) of SWH and compared the results with those of Wang and Swail [4], which were derived from the global hindcast using the NRA wind fields. It was shown that, while showing trend patterns similar to those derived from the global hindcast, the detailed North Atlantic hindcast shows more significant increases in the region off the Canadian east coast in summer and fall, and higher rates of increase in the region northwest of Ireland but less significant changes in the North Sea and in the region off the Scandinavian coast. The homogeneity of the wave data from the detailed North Atlantic hindcast was also assessed in this study. It was concluded that there is no evidence that the kinematic reanalysis of wind fields compromised the homogeneity for improvement in accuracy.

As an extension of the previous studies, here we use different approaches to assess observed SWH changes. We use the trend analysis technique described in Section 2.4 to assess trends in the observed seasonal mean SWH and in the location and scale parameters of seasonal extreme SWH. For the latter, specifically, the following nonstationary GEV models were fitted to the observed seasonal maxima of SWH:

$$\mathrm{GEV}_0(\mu,\ \sigma,\ \xi), \tag{10}$$

$$\mathrm{GEVt}_1(\mu_t = a_0 + \beta_1 t + \varepsilon_t,\ \sigma,\ \xi), \tag{11}$$

$$\mathrm{GEVt}_2(\mu_t = a_0 + \beta_1 t + \varepsilon_t,\ \log(\sigma_t) = b_0 + \beta_2 t + \varepsilon_t,\ \xi), \tag{12}$$

$$\mathrm{GEVt}_3(\mu_t = a_0 + \beta_1 t + \varepsilon_t,\ \log(\sigma_t) = b_0 + \beta_2 t + \theta_2 t^2 + \varepsilon_t,\ \xi), \tag{13}$$

$$\mathrm{GEVt}_4(\mu_t = a_0 + \beta_1 t + \theta_1 t^2 \varepsilon_t,\ \log(\sigma_t) = b_0 + \beta_2 t + \theta_2 t^2 + \varepsilon_t,\ \xi). \tag{14}$$

These GEV models are similar to those described in Section 2.3 except that the covariates here are time variables $t$ and $t^2$, instead of $P_t$ and $G_t$ (the white noise term $\varepsilon_t$ is necessary here because $t$ and $t^2$ are determinate functions and do not represent random variation in the related parameter). Thus, the same likelihood ratio tests were conducted to assess statistical significance of the trend components built in these models.

As a result, we found that changes in the observed seasonal mean SWH are basically linear (except for fall seasonal mean SWH in the North Atlantic where quadratic trends seem not to be negligible in areas off the African coast and the northeast Atlantic), while changes in the observed SWH extremes can be well represented by a linear trend in the location parameter of GEV distribution. Any change in the location parameter can lead to a change in the size or frequency of extreme SWH events. Based on the fitted GEV model:

$$\mathrm{GEV}t_1(\hat{\mu}_t = \hat{a}_0 + \hat{\beta}_1 t,\ \hat{\sigma},\ \hat{\xi}), \tag{15}$$

i.e., a GEV distribution with a time-varying location parameter, we estimated changes in 20-year return values of SWH ($H_{20\text{year}}$) observed in the 25-year period from 1975 to 2000, or, alternatively, the year 2000's return periods of the year 1975's 20-year return values of SWH.

In the North Pacific, changes in winter SWH observed during 1958–1997 are characterized by significant increases in the west–central North Pacific, showing

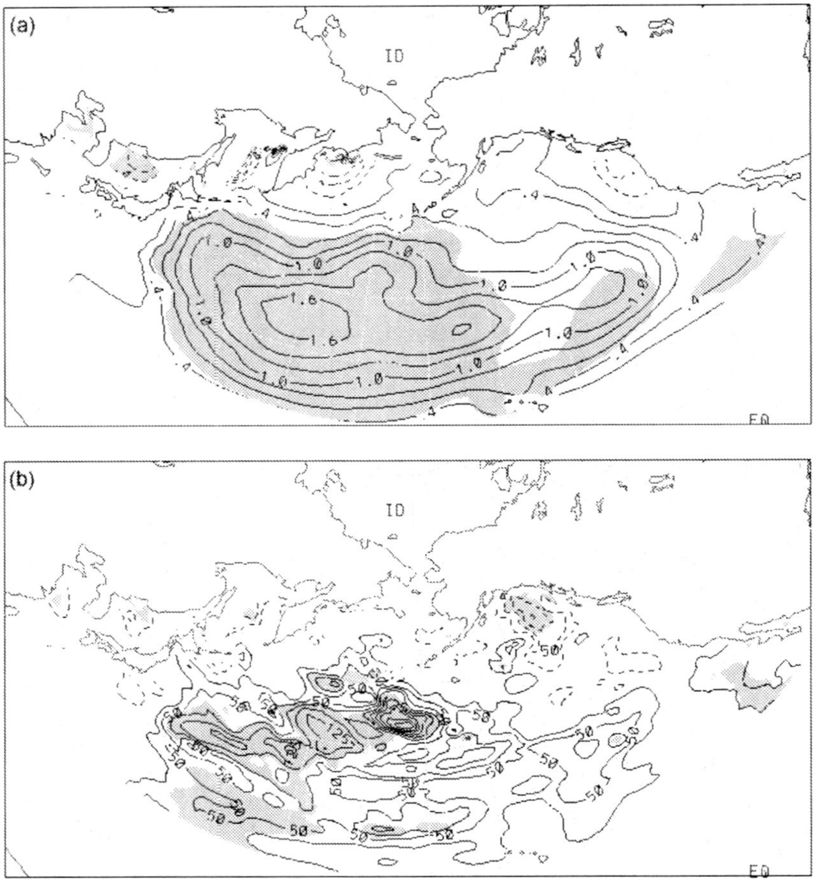

Figure 2: (a) Rates of change (cm/year) in the winter seasonal means of SWH in the North Pacific during 1958–1997. Solid and dashed lines are positive and negative contours, respectively (zero contours are not drawn). Shading indicates areas of significant (5% level) changes. (b) Similar to (a) but for changes in winter 20-year return values of SWH from 1975 to 2000 (2000s–1975s), contoured with an interval of 25 cm.

0.4–1.8 cm/year increases in the seasonal mean SWH (Fig. 2a), as well as 50–250 cm increases in the 20-year return values of SWH in the 25-year period from 1975 to 2000 (Fig. 2b). The 1975's 20-year return values of winter SWH (estimated using the location parameter values as of year 1975) are shown in Fig. 3a. These extreme values of SWH, which occur on average once every 20-year period in the 1975's climate, are expected to occur on average once every 5–10-year period in the 2000's climate (cf. Fig. 3b), according to the estimated linear trend in the location parameter of GEV distribution [eqn (15)]. In fall, changes in the observed SWH are similar, except that the changes are smaller and the area of increase

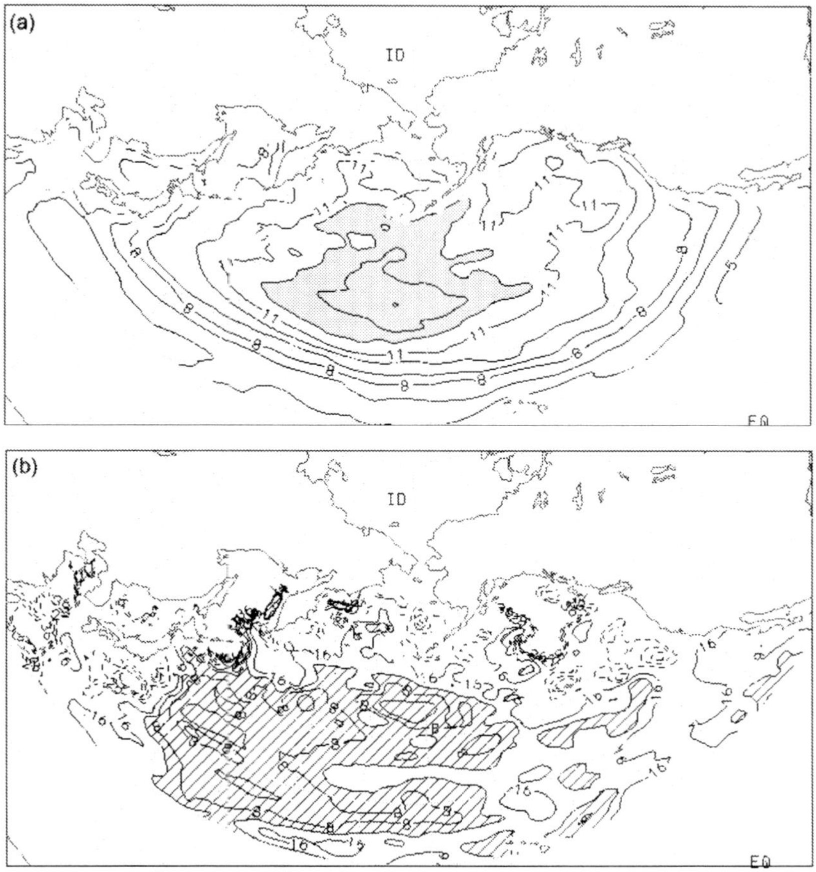


Figure 3: (a) Winter 20-year return values of SWH (in m) estimated for the 1975's climate. (b) 2000's return periods of the 1975's 20-year return values of winter SWH, contoured with an interval of 4 years. Hatching indicates areas where the return period is shorter than 12 years.

shrinks westward to the area west of 170°E (see Fig. 4). In the area of significant changes in extreme SWH (shaded areas in Fig. 4b), the return period of a fixed size of extreme SWH also shortened approximately by half during the 25-year period (1975–2000). In the northwestern Pacific, changes in the mean SWH identified here are comparable with those reported by Yamaguchi and Hadata [28], who produced a 51-year wave hindcast for the northwestern Pacific (approximately 15°N–45°N, 110°E–150°E; using a shallow water wave model driven by the NRA 10-m winds) that shows a strong increasing trend (3–7 mm/year) in the yearly mean SWH in the region south and east of Japan.

As shown in Fig. 5a, significant increases of 1–3 cm/year were estimated for winter mean SWH observed in the northeast Atlantic during 1958–1997, which

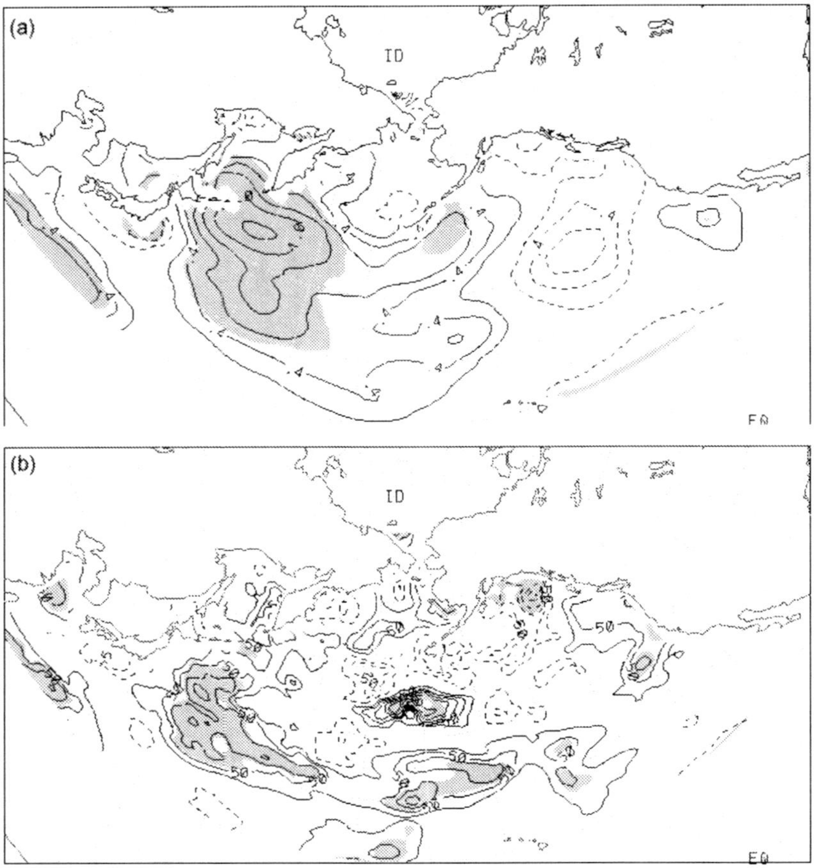

Figure 4: The same as in Fig. 2a and b, respectively, but for the fall seasonal SWH.

are matched by decreases of 0.4–2 cm/year in the midlatitudes. The pattern of changes in winter extreme SWH is similar but noisier. In the northeast Atlantic, the $H_{20\text{year}}$ increased by 50–200 cm during 1975–2000 (cf. Fig. 6a), or the return period of a fixed size of extreme SWH shortened approximately by half: an extreme SWH that occurs on average once every 20-year period in the 1975's climate is expected to occur once every 5–10-year period in the 2000's climate (cf. Fig. 7a).

The major difference between the patterns of change in winter and fall SWH in the North Atlantic is that, in the area off the Canadian east coast, significant increases were found for fall but decreases were identified for winter (cf. Fig. 5–7). However, changes in the northeast Atlantic are smaller in fall than in winter.

In both the North Atlantic and the North Pacific, the changes of winter and fall extreme SWH identified here are consistent with those identified by Wang and Swail [3, 4], although different approaches were taken in the analyses.

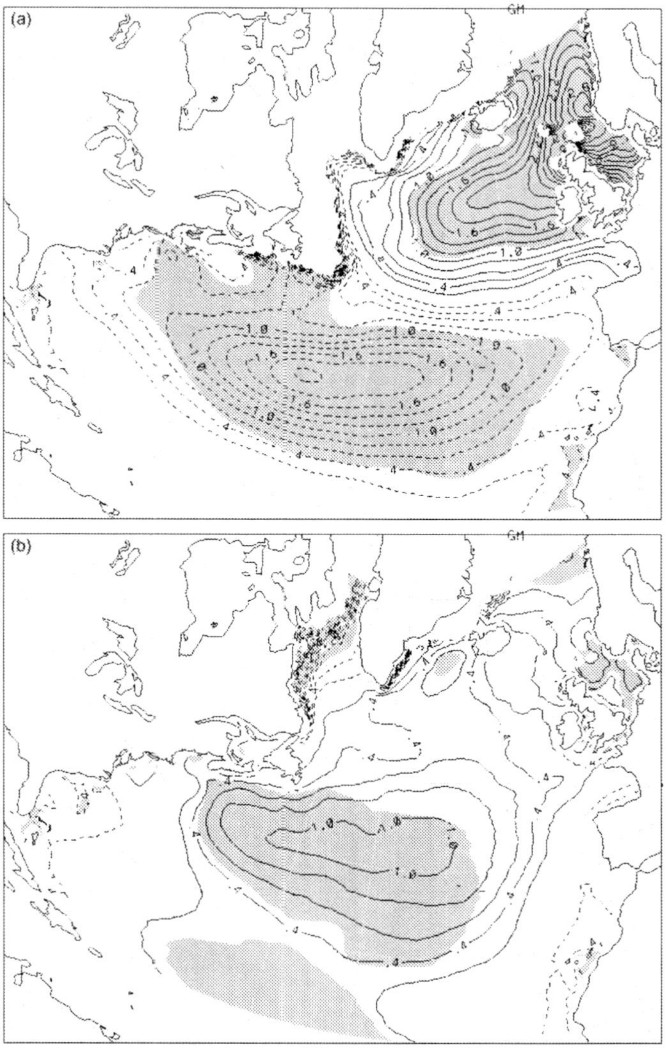


Figure 5: Rates of change (cm/year) in (a) winter and (b) fall mean SWH in the North Atlantic during 1958–1997. Solid and dashed lines are positive and negative contours, respectively (zero contours are not drawn). Shading indicates areas of significant (5% level) changes.

## 4 Climate change scenarios of wave heights

As mentioned before, projections of ocean wave heights are useful for the design and operation of coastal and offshore industries. However, previous studies in this field are limited. The STOWASUS-2100 group [29] carried out two 30-year

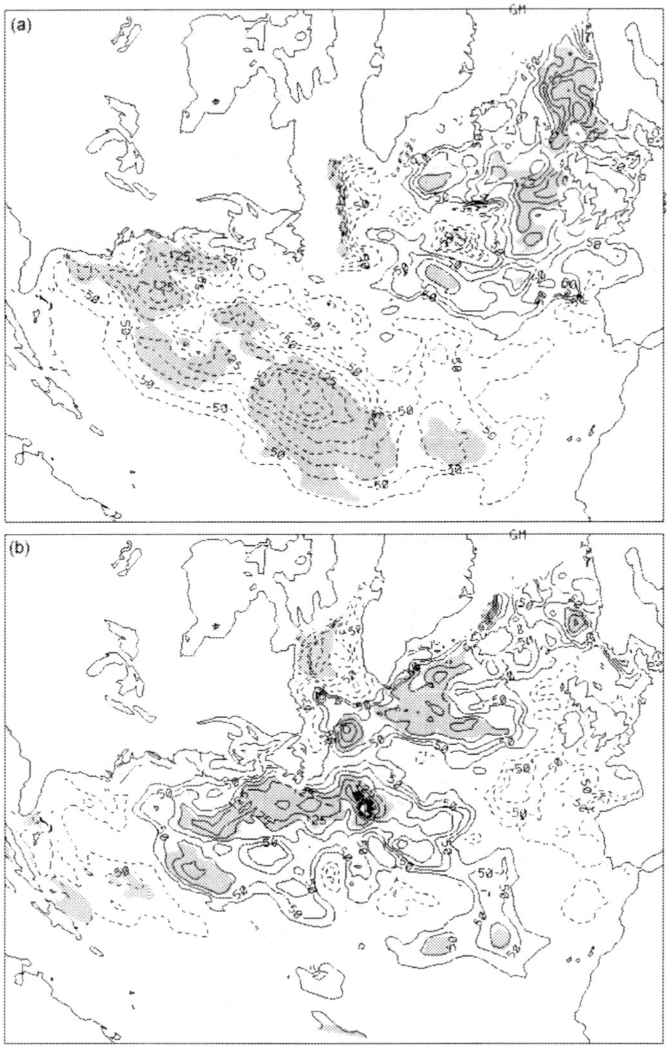


Figure 6: Similar to Fig. 5 but for changes in the (a) winter and (b) fall 20-year return values of SWH from 1975 to 2000 (2000s–1975s), contoured with an interval of 25 cm (zero contours are not drawn).

time-slice experiments for the northeast Atlantic: a control run for the period 1970–1999, and a double $CO_2$ run for the period 2060–2089. The WASA group [19] carried out a similar pair of 5-year time-slice experiments for the North Atlantic as well as producing a statistical projection of future anomalies of intramonthly quantiles of wave heights at Brent and near Ekofisk (in the northern and central North Sea, respectively). Given the intensity of industrial activity in parts of the northern hemisphere oceans, and the intensity of ship traffic throughout the basins,

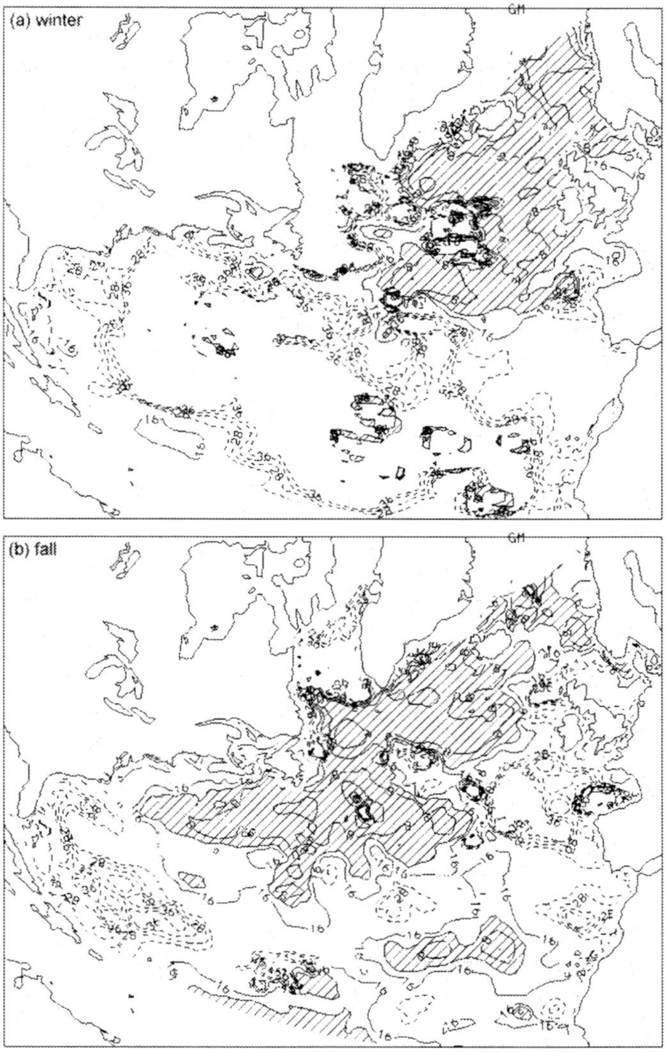


Figure 7: The year 2000's return periods of the year 1975's 20-year return values, contoured with an interval of 4 years. Hatching indicates areas where the return period is shorter than 12 years.

there is a need for projections of possible future climate of wave heights in the North Pacific, and for additional wave heights climate projections that span the entire North Atlantic basin. Recently, Wang *et al.* [11] constructed climate change scenarios of SWH in the North Atlantic by means of RA (for seasonal mean SWH) and nonstationary GEV analysis (for seasonal extreme SWH), using projections of SLP from a coupled climate model (CGCM2). In this study, we extend the wave heights climate change scenarios of Wang *et al.* [11] to the North Pacific basin.

Table 3: The field significance of trends in the projected (with three forcing scenarios) seasonal means (i.e., $h_t$) or in the projected GEV location parameter (i.e., $\hat{\mu}_t$) of seasonal extremes of SWH in the North Atlantic and the North Pacific, as represented by the percentages of grid points where a RM is favored over a relatively simpler RM at the 5% level of significance (Wang *et al.* [11]).

| Forcing scenarios | RMs evaluated | $h_t$ | | | | $\hat{\mu}_t$ | | | |
|---|---|---|---|---|---|---|---|---|---|
| | | NA | | NP | | NA | | NP | |
| | | JFM | OND | JFM | OND | JFM | OND | JFM | OND |
| IS92a | $RM_1$ over $RM_0$ | 67.7 | 26.8 | 82.6 | 66.4 | 59.2 | 54.2 | 79.5 | 67.3 |
| | $RM_2$ over $RM_0$ | 69.5 | 20.0 | 79.7 | 59.9 | 59.3 | 48.8 | 81.4 | 63.9 |
| | $RM_2$ over $RM_1$ | 23.5 | 0.0 | 27.9 | 0.0 | 10.2 | 6.3 | 20.9 | 6.3 |
| A2 | $RM_1$ over $RM_0$ | 61.1 | 60.0 | 81.9 | 69.4 | 61.7 | 55.0 | 78.0 | 67.4 |
| | $RM_2$ over $RM_0$ | 75.7 | 54.5 | 85.0 | 71.7 | 60.3 | 51.2 | 84.0 | 62.5 |
| | $RM_2$ over $RM_1$ | 47.7 | 4.2 | 58.7 | 16.6 | 14.6 | 6.7 | 50.9 | 7.8 |
| B2 | $RM_1$ over $RM_0$ | 45.3 | 6.9 | 68.2 | 51.8 | 51.5 | 32.5 | 62.9 | 54.5 |
| | $RM_2$ over $RM_0$ | 44.3 | 1.0 | 64.0 | 42.3 | 53.6 | 22.7 | 59.2 | 49.6 |
| | $RM_2$ over $RM_1$ | 9.1 | 0.0 | 12.1 | 0.0 | 9.7 | 0.9 | 11.4 | 2.9 |

The scenario construction and analysis procedures are the same as proposed by Wang *et al.* [11], which were also briefly described in Section 2.

Results of the trend analysis (see Section 2.4) on the projected seasonal mean SWH ($h_t$) and on the projected location parameter ($\hat{\mu}_t$) of seasonal extreme SWH (summarized in Table 3) show that changes in both $h_t$ and $\hat{\mu}_t$ are basically nonlinear, especially in winter SWH projected with the strong forcing scenarios (IS92a and A2), and that the nonlinearity of trend appears to be slightly more extensive in the North Pacific than in the North Atlantic (see rows "$RM_2$ over $RM_1$" in Table 3). Therefore, the estimated quadratic trends (i.e., the fitted $RM_2$) for both $h_t$ and $\hat{\mu}_t$ were chosen for use in the estimation of possible future changes in SWH, which are to be described in the subsections below (note that $\hat{\alpha}_2$ is near zero if the trend is basically linear).

## 4.1 Wave height climate change scenarios for the North Atlantic

Wang *et al.* [11] described in detail the climate change scenarios of wave heights in the North Atlantic. For completeness, we briefly summarize the scenarios in this section.

The projected changes of seasonal mean SWH in the 90-year period from 1990 to 2080, as estimated using the fitted $RM_2$:

$$\hat{h}_{tr}(t) = \hat{\alpha}_0 + \hat{\alpha}_1 t + \hat{\alpha}_2 t^2 \tag{16}$$

are partly shown in Fig. 8. Generally, the changes projected under the IS92a and A2 scenarios are larger and more extensive than those under the B2 scenario (Fig. 8a–c),

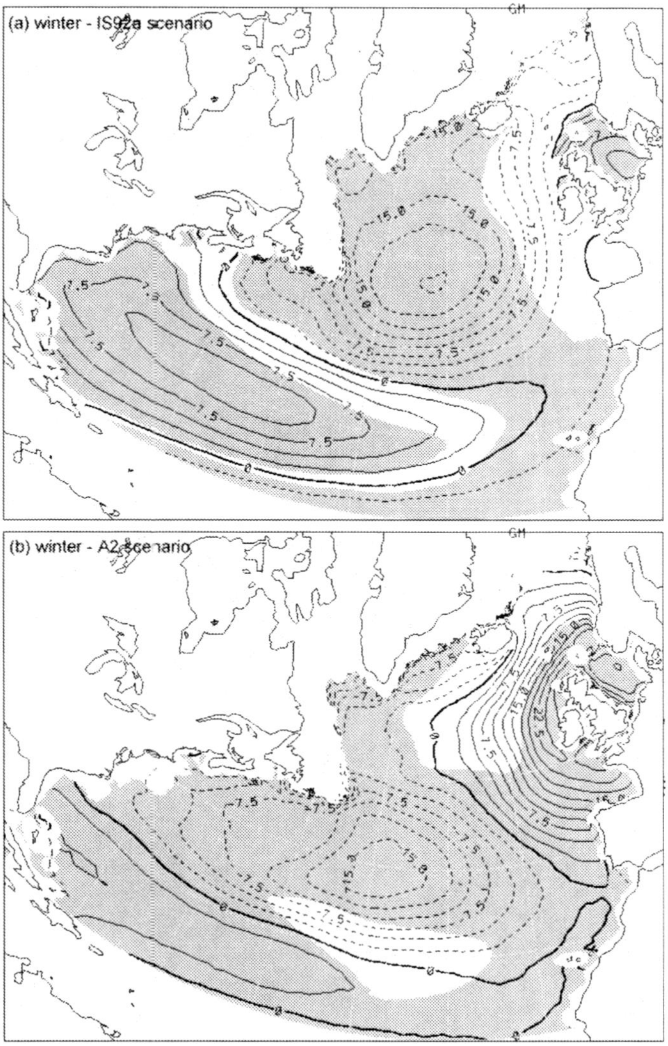


Figure 8: Changes in the seasonal means of SWH in the period from 1990 to 2080 (2080s–1990s), as projected with the indicated forcing scenario. The contour interval is 2.5 cm. Solid and dashed lines are positive and negative contours, respectively. Shading indicates areas of significant quadratic trends (cited from Wang *et al.* [11]).

although the three scenarios have similar patterns of change in both seasons (which feature increases in the northeast and the southwest North Atlantic that are accompanied by decreases in the midlatitudes). The projected changes are generally smaller and less extensive in fall than in winter (cf. Fig. 8b and d, also Table 3).

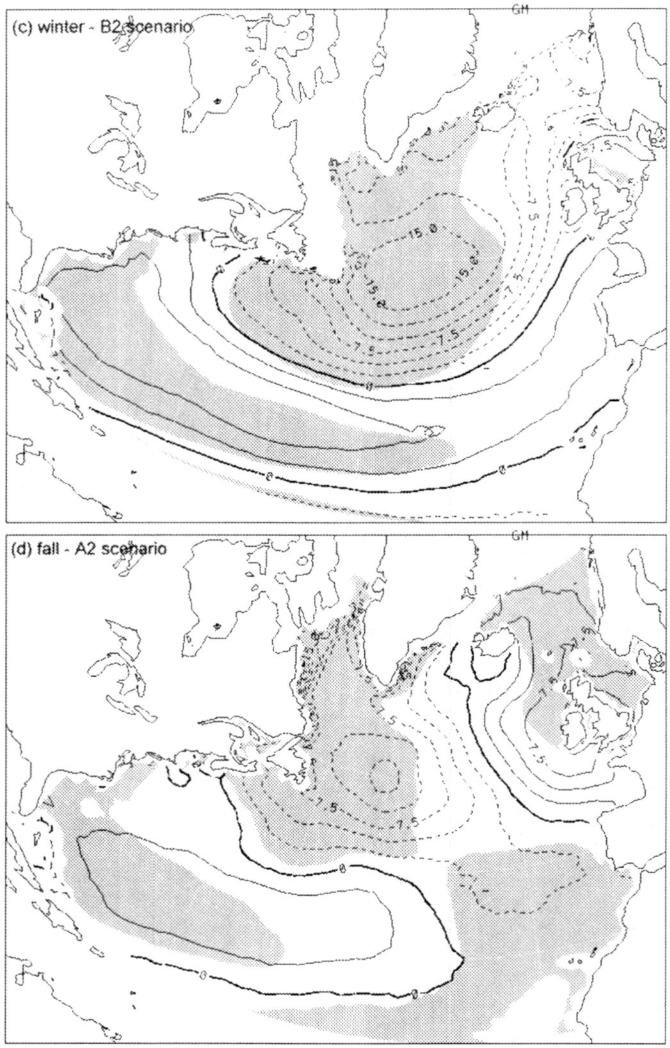


Figure 8 (*continued*)

However, future changes of seasonal mean SWH in the northeast Atlantic (especially in the North Sea) were found to be quite dependant on the forcing conditions. In winter, for example, quadratic trends with decreases in the first few decades of the 1990–2099 period followed by increases were projected with the IS92a scenario, while large and basically linear increases were projected with the A2 scenario and no significant changes with the B2 scenario (Wang *et al.* [11]). In fall, quadratic components of trends were found to be less significant in general, and are basically characterized by faster increases in the late decades than in the first few decades (Wang *et al.* [11]).

The characteristics of the quadratic trends in the projected location parameter of SWH extremes are generally similar to those found in the projected seasonal mean SWH. In the Norwegian Sea and the North Sea, the quadratic trends again feature either faster increases in the late decades than in the first few decades or a decreasing trend in the first few decades followed by an increasing trend (Wang *et al.* [11]). However, the projected changes of fall extreme SWH in the Norwegian Sea appear to be more nonlinear than those projected for the seasonal means under strong forcing (IS92a and A2) scenarios (Wang *et al.* [11]). The estimated quadratic trend of the location parameter

$$\hat{\mu}_{tr}(t) = \hat{\alpha}_0 + \hat{\alpha}_1 t + \hat{\alpha}_2 t^2 \tag{17}$$

and the fitted GEV model

$$\text{GEV}_2(\hat{\mu}_{tr}(t), \hat{\sigma}, \hat{\xi}) \tag{18}$$

were chosen for use in estimating possible future changes in extreme SWH.

Figure 9 shows the changes in 20-year return values of SWH projected for the 90-year period from 1990 to 2080 (the B2 scenario projections are not shown here; they are similar to those of the A2 scenario, except that the significant increases in the northeast Atlantic projected with the A2 scenario are completely absent in the B2 scenario projections). These patterns of change bear substantial similarity to the projected patterns of change in the related seasonal mean SWH (see Figs 8 and 9): Increases of 15–100 cm in the northeast and southwest North Atlantic are now matched by decreases of 15–75 cm in the midlatitudes. However, in the northeast Atlantic, increases in the seasonal extremes are much larger than those in the seasonal means in the fall season but are smaller in the winter season. The largest increases in the extreme SWH in this region are now projected for the fall season rather than for the winter season (see Fig. 9). The between-scenarios differences are also similar, with the A2 (B2) scenario being associated with the greatest (smallest) increases in the northeast Atlantic.

Equivalent to increases in the SWH extremes of a fixed frequency of occurrence are reductions in the average waiting time between extreme SWH events of a fixed size. In the Norwegian Sea (around 65°N), for example, the fall extreme SWH that occurs on average once every 20-year period in the present-day (1990s) climate will occur on average once every 4–12 year period in the climate projected for year 2080 (Wang *et al.* [11]).

## 4.2 Wave height climate change scenarios for the North Pacific

In the North Pacific, as shown in Fig. 10, all the three forcing scenarios again have similar patterns of change in winter mean SWH, which feature a band of large increases (up to 70 cm) in the east–central North Pacific that are accompanied with decreases (up to 30 cm) in the Gulf of Alaska and the southwest North Pacific. However, the changes projected under the strong forcing (IS92a and A2) scenarios are larger and/or more extensive than those under the weaker forcing (B2) scenario.

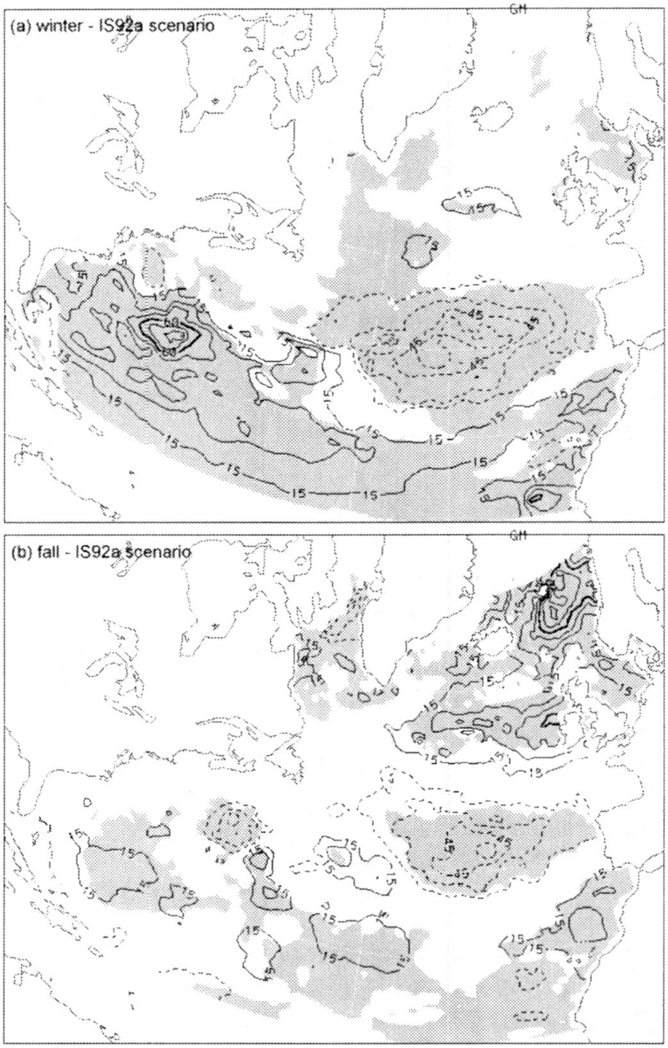


Figure 9: Similar to Fig. 8 but for the projected changes of the winter and fall 20-year return values of SWH, contoured with an interval of 15 cm. Shading indicates areas of significant quadratic trends in the location parameter of the SWH extremes (cited from Wang *et al.* [11]).

For example, the amplitude of increase projected under the B2 scenario is only half of that projected under the IS92a scenario (Fig. 10a and c), and decreases in the region off the Asian coast are apparent only under the strong forcing scenarios (Fig. 10a and b). Although the projected changes in fall mean SWH have similar between-scenarios differences (not shown), these changes are generally much smaller and less extensive than those of winter mean SWH (cf. Table 3) and are

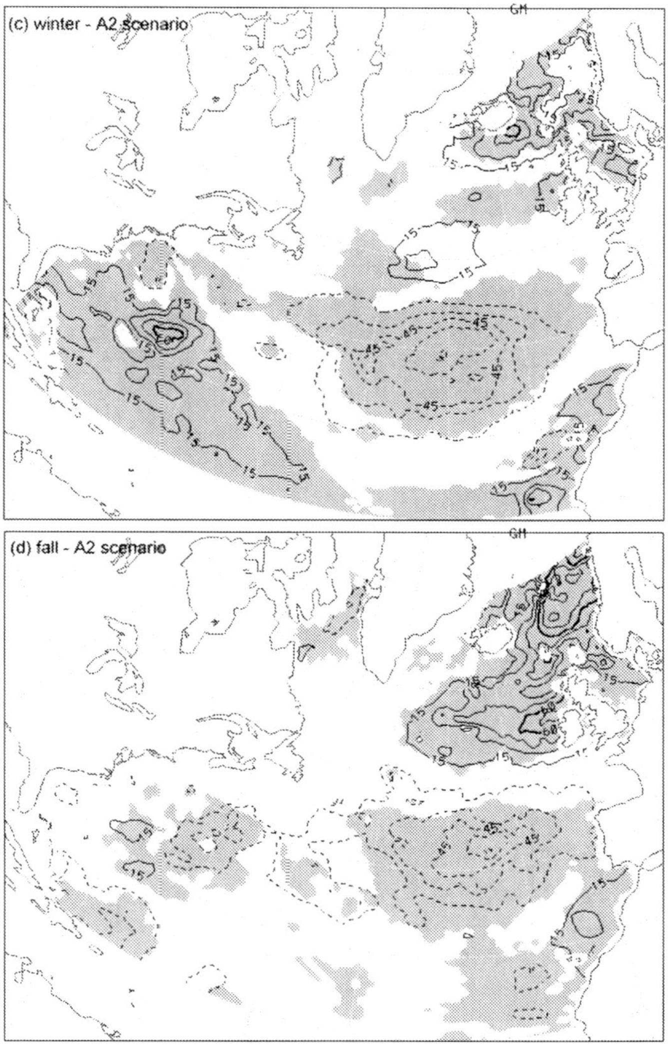


Figure 9 (*continued*)

characterized by increases in the midlatitudes of eastern North Pacific and decreases in the northwest Pacific and in the Gulf of Alaska (Fig. 10d). Note that the center of increase is located closer to the North American west coast in fall than in winter (thus, the region just off the North American west coast has larger increases in fall; cf. Fig. 10b and d). Also, in the southwest North Pacific, small increases were identified in fall but decreases were found in winter.

As mentioned before, nonlinear changes are dominant in wave heights projected for the 21st century, especially in the winter season (JFM; cf. Table 3). In the midlatitudes of eastern North Pacific (around 40°N and 135°W), the nonlinear trend

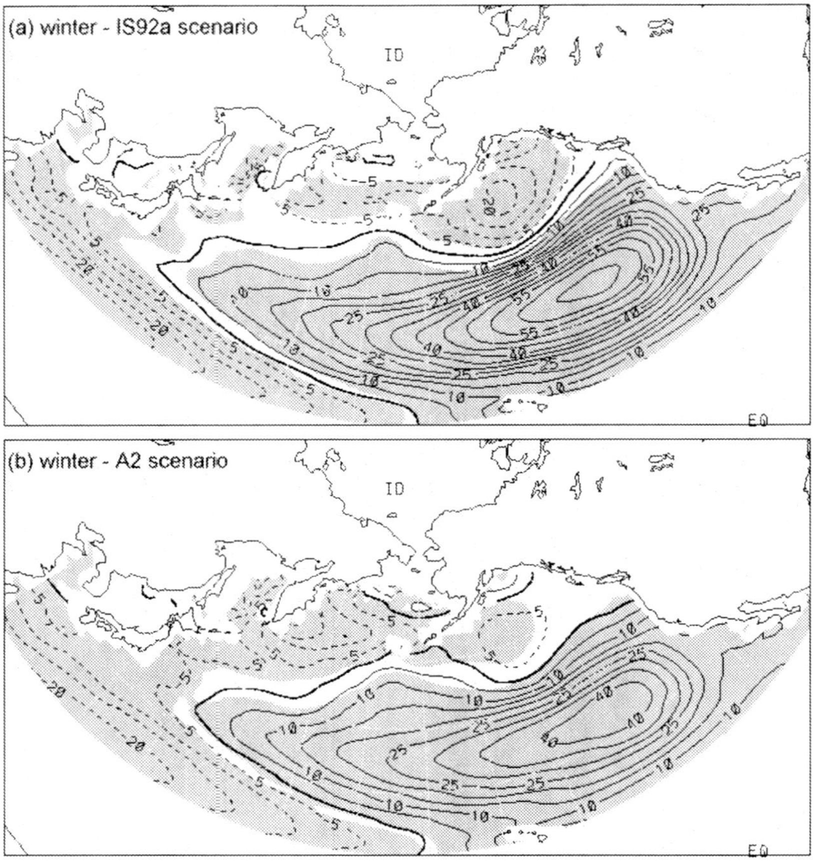


Figure 10: Changes in the winter (a–c) and fall (d) seasonal means of SWH in the North Pacific in the period from 1990 to 2080 (2080s–1990s), as projected with the indicated forcing scenarios. The contour interval is 5 cm. Solid and dashed lines are positive and negative contours, respectively (zero contours are in bold). Shading indicates areas of significant (5% level) quadratic trends.

features an accelerating rate of increase in the second half of the 21st century, which may or may not be preceded by a decreasing trend in the early decades depending on the forcing conditions (cf. left panels of Fig. 11). In the Gulf of Alaska (around 52.5°N and 155°W), the nonlinearity of trend is characterized by an accelerating rate of decrease in the second half of the century (cf. right panels of Fig. 11).

Figure 11 also shows that the characteristics of trends in the projected location parameter of seasonal extreme SWH are similar to those in the projected seasonal mean SWH. Such changes in the location parameter implicate changes in the size of extreme SWH of a fixed occurrence frequency, or, equivalently, in the average waiting time between extreme SWH events of a fixed size.

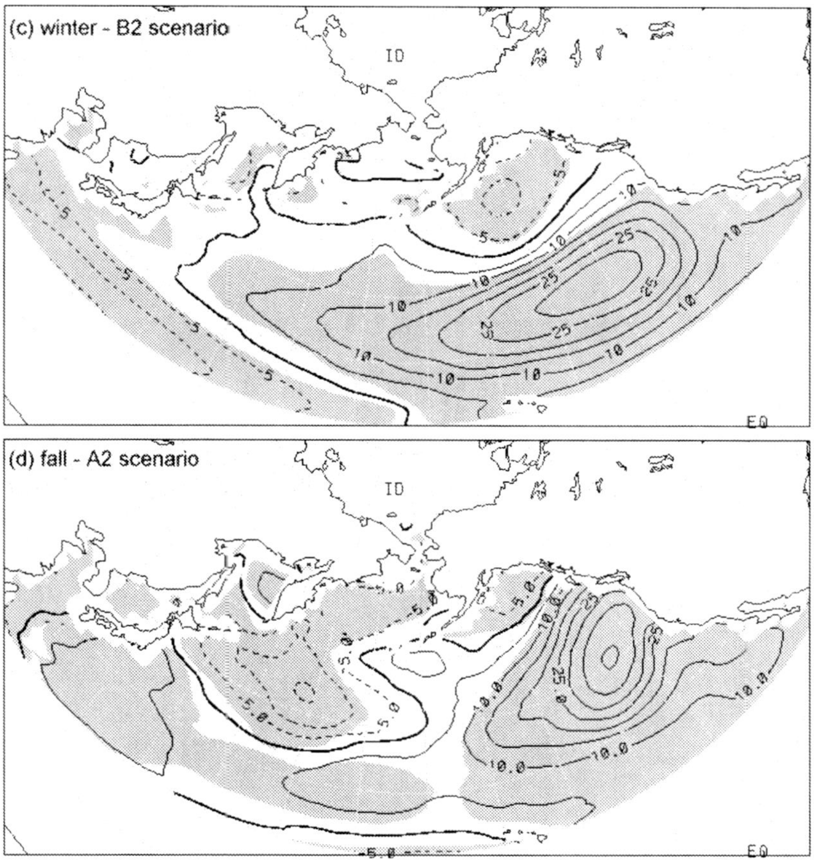


Figure 10 (*continued*)

As shown in Fig. 12, significant increases in the midlatitudes of eastern North Pacific are the dominant feature of the projected changes in the 20-year return values of SWH. The increases range up to about 130 cm in the 90-year period from 1990 to 2080. Decreases are much less apparent in the projected seasonal extremes of SWH than in the corresponding seasonal means (cf. Figs 10 and 12), although the extremes and the means share similar patterns of changes and similar between-scenarios differences. The region near shore the North American west coast was also projected to have larger increases in fall extreme SWH than in winter (cf. Fig. 12b and d).

In terms of changes in the average waiting time between extreme SWH events of a fixed size, Fig. 13 shows that in the midlatitudes of eastern North Pacific an extreme SWH that occurs on average once every 20-year period in the present-day (1990's) climate is projected to occur on average once every 4–12 year period in the 2080's climate under the A2 forcing scenario. In other words, the average waiting time between extreme SWH events of a fixed size was projected to shorten

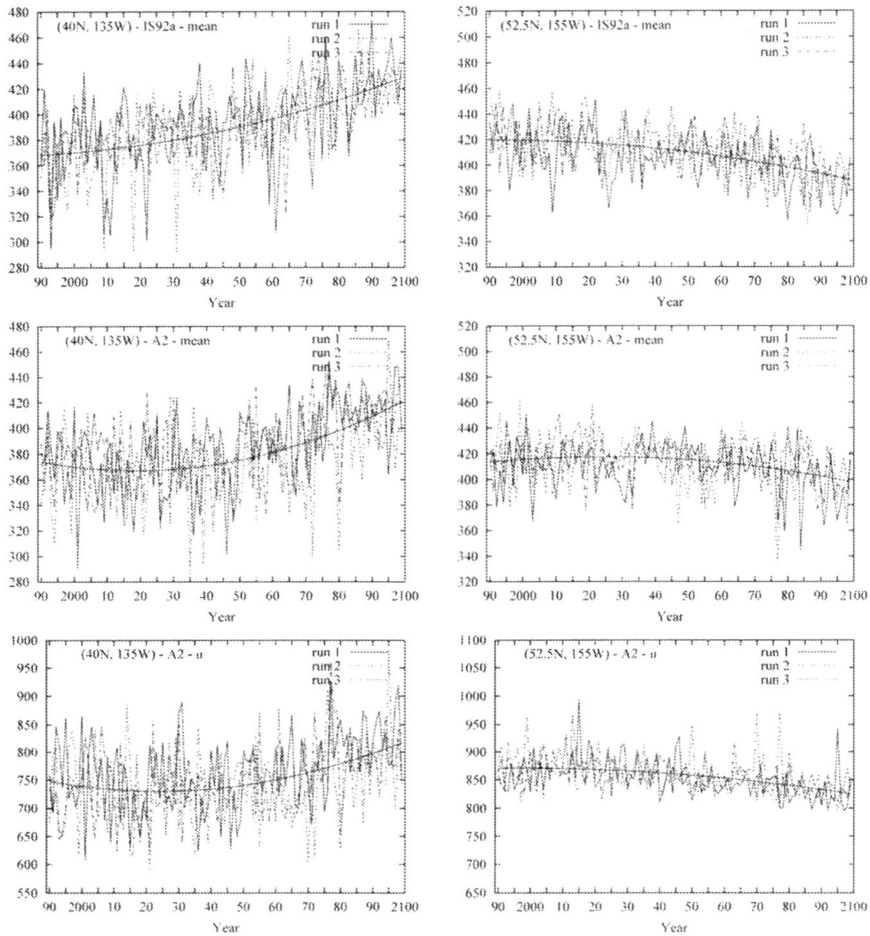


Figure 11: Trends in the winter seasonal mean SWH (upper and middle panels) and in the location parameter of winter SWH extremes (lower panels), as projected with the IS92a and A2 forcing scenarios for the indicated grid points in eastern North Pacific (each scenario projection has an ensemble of three runs). The unit of the vertical axis is centimeter.

approximately by half in the 90-year period from 1990 to 2080, especially in the region just off the North American west coast in fall (cf. Fig. 13b).

## 5 Concluding remarks

We have reviewed and extended previous studies on historical and possible future changes of SWH in the North Atlantic and the North Pacific, separately.

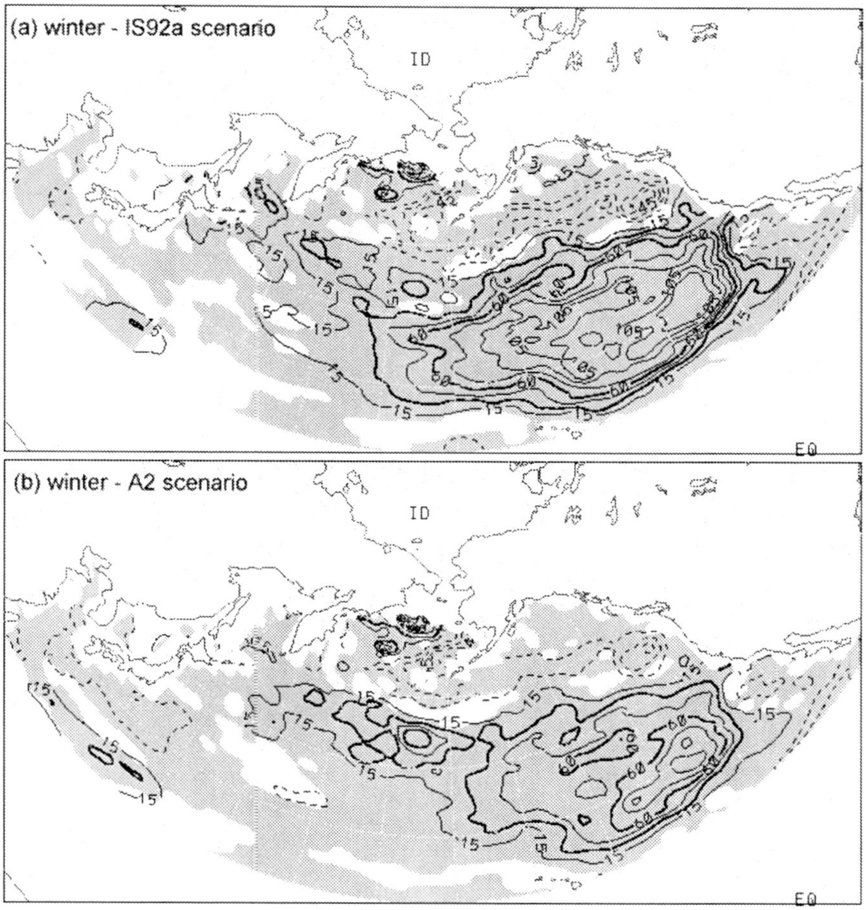


Figure 12: Changes in the winter (a–c) and fall (d) 20-year return values of SWH in the North Pacific in the period from 1990 to 2080 (2080s–1990s), as projected with the indicated forcing scenarios. The contour interval is 15 cm. Solid and dashed lines are positive and negative contours, respectively (zero contours are not drawn). Shading indicates areas of significant (5% level) quadratic trends in the location parameter of SWH extremes.

Two 40-year wave hindcasts that are the best to date were used as observed waves. The SLP–SWH relationships observed in the 40-year period were represented by regression models and used to project future changes of seasonal mean and extreme SWH under the possible future atmospheric circulation (SLP) conditions as projected by CGCM2 with three different forcing scenarios.

In the 40-year period, changes of SWH observed in both the North Atlantic and the North Pacific were found to be statistically significant and basically linear.

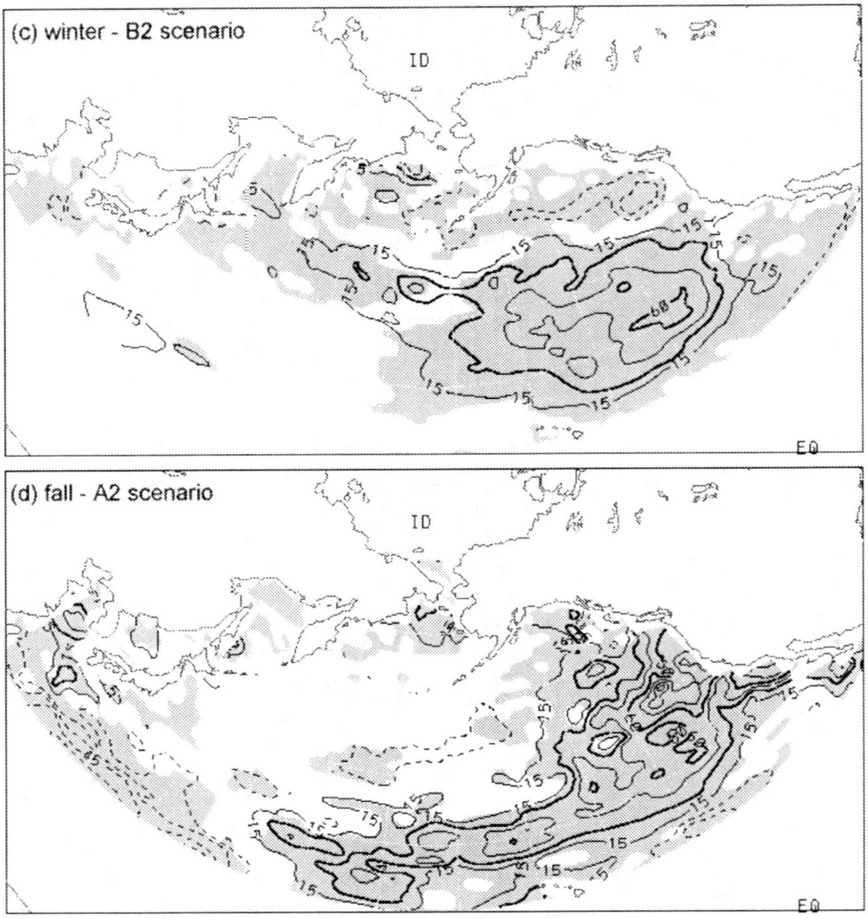


Figure 12 (*continued*)

The changes of winter SWH in the North Atlantic are characterized by increases in the northeast Atlantic matched by decreases in the midlatitudes, which were found to be closely associated with changes in the NAO. Significant increases in the area off the Canadian east coast are the dominant feature of changes in fall SWH in the North Atlantic. The changes of both winter and fall SWH in the North Pacific are characterized by a center of significant increases in the west–central North Pacific (Wang and Swail [4]), which were found to be connected with a deeper and more eastward extended Aleutian low or, equivalently, with a southward dip and eastward extension of the storm track (Graham and Diaz [25]).

It was shown that stronger (weaker) GHG forcing conditions generally lead to larger (smaller) changes in ocean wave heights, and that in some regions (e.g., the North Sea) the ocean wave climate responds very differently to the different forcing scenarios. We may expect to see an imprint of the anthropogenic forcing in the ocean wave climate change in the coming decades.

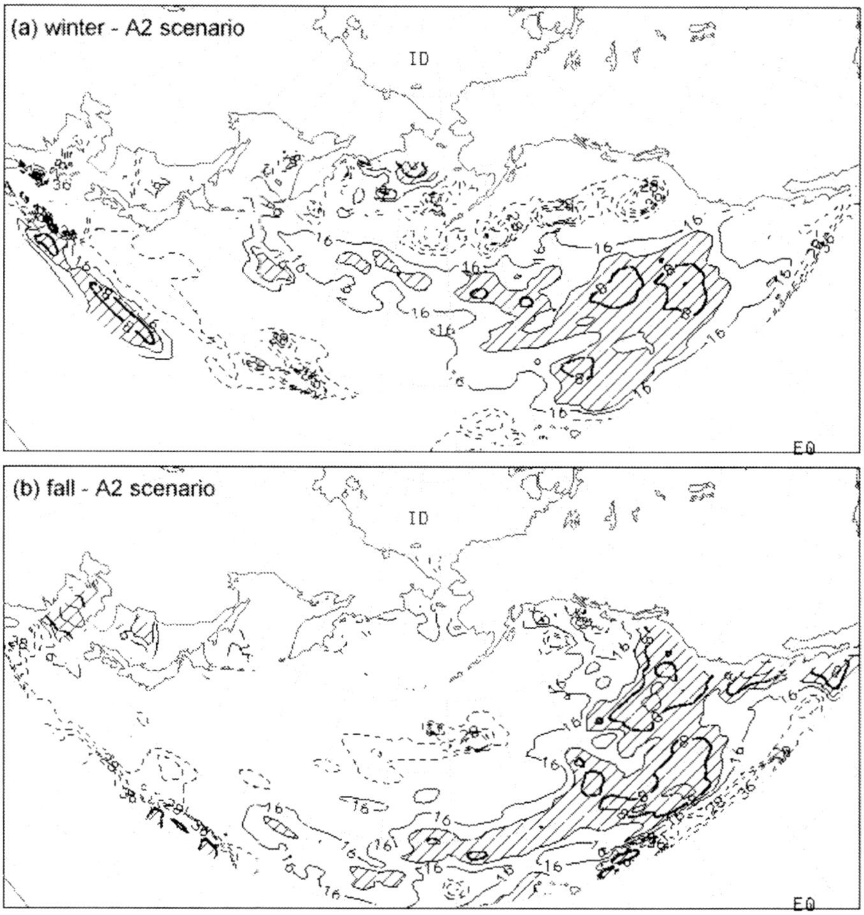


Figure 13: Future (2080's) return periods of the 1990's 20-year return values of SWH, as estimated using the GEV parameters values projected for year 2080 under the A2 forcing scenario. The contour interval is 4 years. Solid and dashed contours indicate return periods shorter or longer than 20 years, respectively (20-year contours not drawn). Hatching indicates values $\leq 12$ years.

It was also shown that changes in both the observed and the projected seasonal extremes of SWH can well be represented by changes in the location parameter of the extreme value distribution. Global warming may result in changes in the location parameter, eventually leading to changes in the size and frequency of extreme wave height events. For example, in the subtropical eastern North Pacific, the average waiting time between extreme SWH events of a fixed size was projected to shorten approximately by half in the 90-year period from 1990 to 2080, especially in the region near shore the North American west coast in fall; and a similar change was also projected for the Norwegian Sea (around 65°N; Wang *et al.* [11]). Such significant changes will have an impact on the life span of marine and coastal infrastructure in these areas. The possible changes in future wave extremes should be taken into account in the design/planning and operation of coastal and offshore industries.

Note that, in both the North Atlantic and North Pacific, the patterns of the projected SWH changes in the 21st century differ to some extent from those of the observed SWH changes during the past half century. In the North Atlantic, such differences are especially remarkable in fall, featuring observed increases in the region off the Canadian east coast versus decreases or no changes in the related projections (cf. Figs 5–9). In the North Pacific, the differences are noticeable in both seasons and are characterized by an eastward shift of the center of increase, with the observed increases centered in the western North Pacific while the projected increases centered in the eastern North Pacific (cf. Figs 2–4, 10, and 12). In other words, trends of SWH in the projected warmer climate are not just a simple continuation of the trends observed in the last half century.

For the North Atlantic sector, it has been shown that global warming is associated with more frequent occurrence of the positive phase of NAO (or less frequent occurrence of the extreme negative phase of NAO; Monahan *et al.* [30]) on the one hand, and with increases of wave heights in the northeast Atlantic on the other hand. This implicates that the projected wave height increases in the northeast Atlantic are associated with the anthropogenic changes that affect the NAO (Wang *et al.* [11]). In the projected warmer climate, decreases of SLP over the northeast Atlantic are accompanied with increases of SLP over the midlatitudes, which result in a pressure gradient increase and more frequent and stronger westerly winds that affect northern European coasts (Wang *et al.* [11]).

For the North Pacific sector, as shown in Fig. 14, the CGCM2 simulations project an intensification and southeastward expansion of the Aleutian low for the warmer climate, which is particularly remarkable in winter but also noticeable in fall. Consistent with these SLP climate changes are the projected increases of SWH in the midlatitudes of eastern North Pacific (cf. Figs 10, 12, and 14). Some other transient simulations with coupled atmosphere models have also shown an intensification and expansion of the Aleutian low during the latter decades of the 20th century and early decades of the 21st century (Roeckner *et al.* [31]; Meehl and Washington [32]; Cai and Whetton [33]; Graham and Diaz [25]).

Changes in SLP and SWH are associated with changes in the position and intensity of the storm tracks in the northern hemisphere, which have been seen in the observed data and can be expected in the projected warmer climate. One of our ongoing projects is a detailed analysis of possible future changes of northern storm tracks as projected by CGCM2, which is to be published soon. Also, we have initiated a study to assess the uncertainty of wave height change scenarios due to different numerical wave hindcasts and different climate models' projections of SLP.

Finally, note that the basic assumption for the climate change scenarios construction using empirical based techniques is that a relationship (e.g., the SLP–SWH

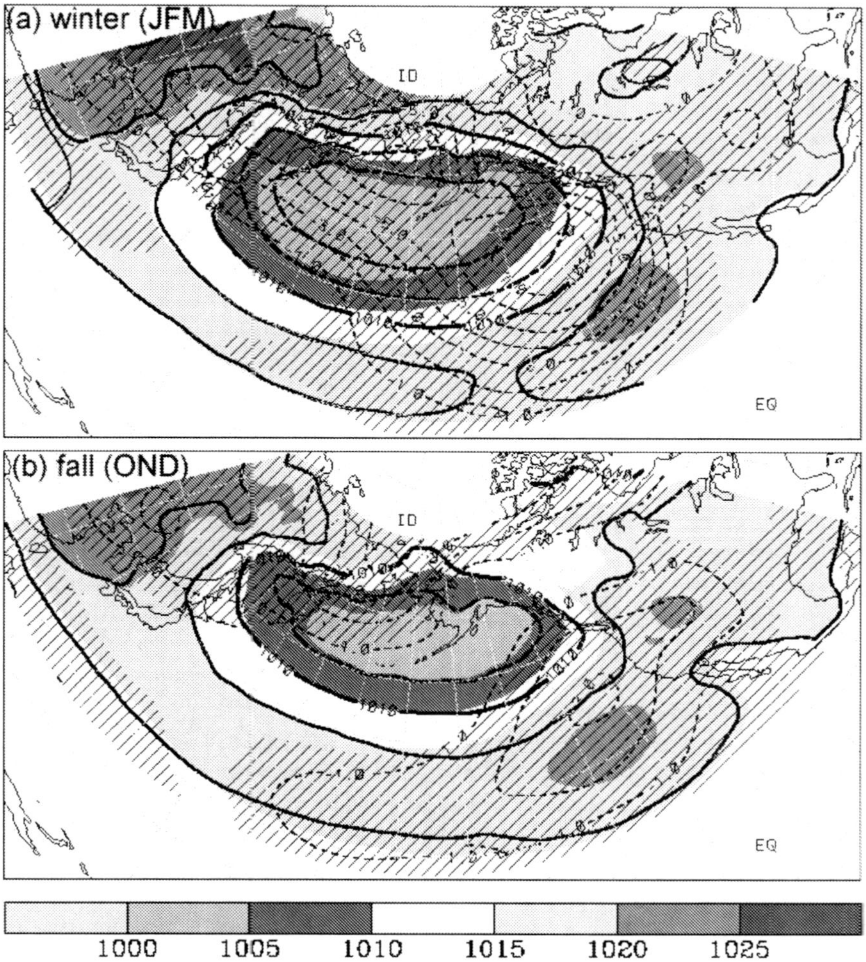


Figure 14: The climate (30-year mean) field of winter and fall SLP for the period of 1961–1990 (1970s; shaded contour map) and of 2070–2099 (2080s; thick-line contour map; contour interval 5 hPa). The thin-line contours show the difference between the two climates (2080s–1970s; contour interval 1 hPa; dashed/solid thin-lines indicate decreases/increases; zero contours not drawn); hatching indicates areas of significant (at 5% level) climate change.

relationship in this study) developed for the present-day climate will continue to hold under the different forcing conditions of possible future climates. In general, empirical based techniques cannot account for possible systematic changes in regional forcing conditions or feedback processes, although they are economical and practical. We should keep in mind the various sources of uncertainty related to climate scenario construction when dealing with climate change scenarios.

## References


[1] IPCC, *Climate Change 2001: The Scientific Basis. Contribution of Working Group I to the Third Assessment Report of the Intergovernmental Panel on Climate Change*, eds J.T. Houghton, Y. Ding, D.J. Griggs, M. Noguer, P.J. van der Linden, X. Dai, K. Maskell & C.A. Johnson, Cambridge University Press: Cambridge and New York, 2001.

[2] IPCC, *Special Report on Emissions Scenarios*, Cambridge University Press: Cambridge, 2000.

[3] Wang, X.L. & Swail, V.R., Trends of Atlantic wave extremes as simulated in a 40-year wave hindcast using kinematically reanalyzed wind fields. *Journal of Climate*, **15(9)**, pp. 1020–1035, 2002.

[4] Wang, X.L. & Swail, V.R., Changes of extreme wave heights in northern hemisphere oceans and related atmospheric circulation regimes. *Journal of Climate*, **14**, pp. 2204–2221, 2001.

[5] Kalnay, E., Kanamitsu, M., Kistler, R., Collins, W., Deaven, D., Gandin, L., Iredell, M., Saha, S., White, G., Woollen, J., Zhu, Y., Chelliah, M., Ebisuzaki, W., Higgins, W., Janowiak, J., Mo, K.C., Ropelewski, C., Wang, J., Leetmaa, A., Reynolds, R., Jenne, R. & Joseph, D., The NCEP/NCAR 40-year reanalysis project. *Bulletin of the American Meteorological Society*, **77**, pp. 437–471, 1996.

[6] Flato, G.M. & Boer, G.J., Warming asymmetry in climate change simulations. *Geophysical Research Letters*, **28**, pp. 195–198, 2001.

[7] Swail, V.R. & Cox, A.T., On the use of NCEP/NCAR reanalysis surface marine wind fields for a long term North Atlantic wave hindcast. *Journal of Atmospheric and Oceanic Technology*, **17**, pp. 532–545, 2000.

[8] Cox, A.T. & Swail, V.R., A global wave hindcast over the period 1958–1997: validation and climate assessment. *Journal of Geophysical Research* **106(C2)**, pp. 2313–2329, 2001.

[9] Cardone, V.J., Pierson, W.J. & Ward, E.G., Hindcasting the directional spectrum of hurricane-generated waves. *Journal of Petroleum Technology*, **28**, pp. 385–394, 1976.

[10] Boer, G.J., Flato, G., Reader M.C. & Ramsden, D., A transient climate change simulation with greenhouse gas and aerosol forcing: experimental design and comparison with the instrumental record for the twentieth century. *Climate Dynamics*, **16**, pp. 405–425, 2000.

[11] Wang, X.L., Zwiers F.W. & Swail, V.R., North Atlantic ocean wave climate change scenarios for the 21st century. *Journal of Climate*, **17**, pp. 2368–2383, 2004.

[12] Wang, X.L. & Zwiers, F.W., Using redundancy analysis to improve dynamical seasonal mean 500 hPa geopotential forecasts. *International Journal of Climatology*, **21**, pp. 637–654, 2001.

[13] von Storch, H. & Zwiers, F.W., *Statistical Analysis in Climate Research*, Cambridge University Press: Cambridge, 1999.

[14] Tyler, D.E., On the optimality of the simultaneous redundancy transformations. *Psychometrika*, **47(1)**, pp. 77–86, 1982.

[15] Bacon, S. & Carter, D.J.T., A connection between mean wave height and atmospheric pressure gradient in the North Atlantic. *International Journal of Climatology*, **13**, pp. 423–436, 1993.

[16] Bacon, S. & Carter, D.J.T., Wave climate changes in the North Atlantic and North Sea. *International Journal of Climatology*, **11**, pp. 545–558, 1991.

[17] Bouws, E., Jannink, D. & Komen, G.J., The increasing wave height in the North Atlantic ocean. *Bulletin of the American Meteorological Society*, **77**, pp. 2275–2277, 1996.

[18] Kushnir, Y., Cardone, V.J., Greenwood, J.G. & Cane, M.A., The recent increase in North Atlantic wave heights. *Journal of Climate*, **10**, pp. 2107–2113, 1997.

[19] WASA Group, Changing waves and storms in the Northeast Atlantic? *Bulletin of the American Meteorological Society*, **79(5)**, pp. 741–760, 1998.

[20] Gunther, H., Rosenthal, W., Stawarz, M., Carretero, J.C., Gomez, M., Lozano, I., Serano, O. & Reistad, M., The wave climate of the Northeast Atlantic over the period 1955–1994: the WASA wave hindcast. *Global Atmosphere-Ocean System*, **6**, pp. 121–163, 1998.

[21] Allan, J.C. & Komar, P., Are ocean wave heights increasing in the eastern North Pacific? *EOS*, **81(47)**, pp. 561, 566–567, 2000.

[22] Gibson, R., Kallberg, P. & Uppala, S., The ECMWF Reanalysis (ERA) Project, *ECMWF Newsletter*, **73**, pp. 7–17, 1996.

[23] Sterl, A., Komen, G.J. & Cotton, P.D., Fifteen years of global wave hindcasts using winds from the European Centre for Medium-Range Weather Forecasts reanalysis: Validating the reanalyzed winds and assessing the wave climate. *Journal of Geophysical Research*, **103**, pp. 5477–5492, 1998.

[24] Graham, N.E., Strange, R.R. & Diaz, H.F., Intensification of North Pacific winter cyclones 1948–1998: impacts on California wave climate. *Preprints of the 7th Int. Workshop on Wave Hindcasting and Forecasting*, Banff, Alberta, Canada, October 21–25, 2002, pp. 60–69, 2002.

[25] Graham, N.E. & Diaz, H.F., Evidence for intensification of North Pacific winter cyclones since 1948. *Bulletin of the American Meteorological Society*, **82(9)**, pp. 1869–1893, 2001.

[26] Caires, S., Sterl, A., Bidlot, J.-R., Graham, N. & Swail, V.R., Climatological assessment of reanalysis ocean data. *Preprints of the 7th Int. Workshop on Wave Hindcasting and Forecasting*, Banff, Alberta, Canada, October 21–25, 2002, pp. 1–12, 2002.

[27] Carter, D.J.T. & Draper, L., Has the north east Atlantic become rougher? *Nature*, **332**, p. 494, 1988.

[28] Yamaguchi, M. & Hatada, Y., 51-year wave hindcast and analysis of wave height climate trend on the northwestern Pacific Ocean. *Preprints of the 7th Int. Workshop on Wave Hindcasting and Forecasting*, Banff, Alberta, Canada, October 21–25, 2002, pp. 48–59, 2002.

[29] STOWASUS Group, *STOWASUS-2100 (Regional Storm, Wave and Surge Scenarios for the 2100 Century)*, Final report (http://www.dmi.dk/pub/STOWASUS-2100/), 2001.

[30] Monahan, A.H., Fyfe, J.C. & Flato, G.M., A regime view of northern hemisphere atmospheric variability and change under global warming. *Geophysical Research Letters*, **27(8)**, pp. 1139–1142, 2000.

[31] Roeckner, E., Oberhuber, J.M., Bacher, A., Christoph, M. & Kircher, I., ENSO variability and atmospheric response in a global coupled atmosphere-ocean GCM. *Climate Dyn.*, **12**, pp. 737–754, 1996.

[32] Meehl, G.A. & Washington, W.M., El Nino-like climate change in a model with increased atmospheric $CO_2$ concentrations. *Nature*, **382**, pp. 56–60, 1996.

[33] Cai, W. & Whetton, P.H., Evidence for a time-varying pattern of greenhouse warming in the Pacific Ocean. *Geophysical Research Letters*, **27**, pp. 2577–2580, 2000.

# Index